SpringerBriefs in Architectural Design and Technology

Series Editor

Thomas Schröpfer, Architecture and Sustainable Design, Singapore University of Technology and Design, Singapore, Singapore

Indexed by SCOPUS

Understanding the complex relationship between design and technology is increasingly critical to the field of Architecture. The *Springer Briefs in Architectural Design and Technology* series provides accessible and comprehensive guides for all aspects of current architectural design relating to advances in technology including material science, material technology, structure and form, environmental strategies, building performance and energy, computer simulation and modeling, digital fabrication, and advanced building processes. The series features leading international experts from academia and practice who provide in-depth knowledge on all aspects of integrating architectural design with technical and environmental building solutions towards the challenges of a better world. Provocative and inspirational, each volume in the Series aims to stimulate theoretical and creative advances and question the outcome of technical innovations as well as the far-reaching social, cultural, and environmental challenges that present themselves to architectural design today. Each brief asks why things are as they are, traces the latest trends and provides penetrating, insightful and in-depth views of current topics of architectural design. *Springer Briefs in Architectural Design and Technology* provides must-have, cutting-edge content that becomes an essential reference for academics, practitioners, and students of Architecture worldwide.

More information about this series at http://www.springer.com/series/13482

Kevin Ka-Lun Lau · Zheng Tan ·
Tobi Eniolu Morakinyo · Chao Ren

Outdoor Thermal Comfort in Urban Environment

Assessments and Applications in Urban Planning and Design

 Springer

Kevin Ka-Lun Lau
Institute of Future Cities
The Chinese University of Hong Kong
Shatin, Hong Kong

Zheng Tan
Department of Building and Real Estate
The Hong Kong Polytechnic University
Hong Kong, Hong Kong

Tobi Eniolu Morakinyo
School of Geography
University College Dublin
Dublin, Ireland

Chao Ren
Faculty of Architecture
The University of Hong Kong
Pok Fu Lam, Hong Kong

ISSN 2199-580X ISSN 2199-5818 (electronic)
SpringerBriefs in Architectural Design and Technology
ISBN 978-981-16-5244-8 ISBN 978-981-16-5245-5 (eBook)
https://doi.org/10.1007/978-981-16-5245-5

This Springer imprint is published by the registered company Springer Nature Singapore Pte Ltd.
The registered company address is: 152 Beach Road, #21-01/04 Gateway East, Singapore 189721, Singapore

Foreword by Prof. Edward Ng

When I was a boy living in Hong Kong, my family lived in an ageing tenement block of a few storeys' height. We had a small fan, but dad used to occupy it. At night, I needed to dress down to my underpants and slept in the steel cage outside our living room window together with a few pot plants. A few steel bars supported and surrounded me. I had no mattress, and it was breezy and cool. I could look up to the starry sky and look down at people walking on the street. If it rained, it was even cooler and more comfortable. This was my first lesson on thermal comfort. It was only much later when I did my architecture degree that I started to understand the six metrics of thermal comfort: air temperature, radiant temperature, humidity, airspeed, clothing, and metabolic rate. Time flies, my mum still lives in the same tenement block. It is now surrounded by tall tower buildings. At night, the air outside is warmer than it used to be 50 years ago. I can hardly see the sky from my steel cage. I can no longer feel the breezes. The wall opposite reflects light and heat. We now turn on our air conditioners to sleep.

As cities urbanise, we create thermal comfort problems in our built environment. We need to understand better the characteristics of the problem we have made. More importantly, we need to know how to prevent and mitigate them. Although Givoni's *Man, Climate and Architecture* (1969) is still a must-read, Dr. Lau's book that focuses on sub-tropical high-density urban living has advanced the scholarly field of the subject matter. The close linkage between thermal comfort and the urban environment allows better design and planning ideas to be developed. Furthermore, a better appreciation of the dynamic responses of pedestrian thermal comfort enables a more heterogeneous, diverse, flexible, and inclusive urban placemaking. As to mitigation measures, urban form and urban greening have been meticulously studied. These understandings are further integrated using physiological equivalent temperature (PET) to be the basis of an urban climatic mapping system for planners.

I wonder what would have happened to one's feelings of the steel cage outside my mum's living room if Dr. Lau's book were to be available before all these tall tower blocks went up. Would planners and architects be more aware and better equipped to design better? And would my grandson be able to sleep comfortably in my steel cage and experience the starry sky the way I had enjoyed it?

Edward Ng
Yao Ling Sun Professor of Architecture
The Chinese University of Hong Kong
Shatin, Hong Kong

Acknowledgements

Before anything else, I would like to express my gratitude and appreciation to my Ph.D. supervisor, Prof. Edward Ng, for giving me different opportunities in my research career. He has been my mentor and role model since I first met him in 2008. I remember what he first told me to do for my Ph.D. study is to read something else, anything but my research field. I was puzzled because I thought I needed to pick up the knowledge required for my Ph.D. study as soon as I could. However, I feel lucky that I did so as I have learned a lot from "something else" and I understand that the problems we face in cities have no single-way solutions. We need to work with different parties to make our cities better. He also taught me to enjoy my work so that my work is enjoyable to not only scholars, but also different people who learn about it. I do enjoy working with him, especially with the wine in his office.

I would also like to thank my co-authors who contribute to the chapters, Dr. Tanya Tan, Dr. Tobi Morakinyo, and Dr. Chao Ren. They have been my research collaborators who taught me a lot throughout my research career. I had the privilege to work with plenty of top-class researchers from different research teams, which provides me numerous possibilities in my work and life.

My interdisciplinary research would not have been possible without the directors I met in the Chinese University of Hong Kong. I would like to thank Prof. Jean Woo, Prof. Yee Leung, Prof. Tung Fung, and Prof. Gabriel Lau, who offered me the opportunities to work with their teams and broaden my horizon in making our cities healthier and more liveable. Working with different people has been challenging but rewarding since I have the chance to rethink and improve my research.

I would like to acknowledge the funding support from the Vice-Chancellor Discretionary Fund of the Chinese University of Hong Kong for Chaps. 2–6, and the Research Grant Council, Hong Kong, for Chap. 2 (General Research Fund, Grant No.: 14643816), Chap. 3 (General Research Fund, Grant No.: 14629516), and Chap. 7 (General Research Fund, Grant No.: 14611015). The research studies would not have been successfully completed without their financial and administrative support.

I would like to acknowledge my assistants, Ms. Mona Chung and Ms. Cheryl Yung, who helped me with the research tasks involved in the book and offered me a helping hand whenever I need.

I am also grateful to my family for their unconditional support to my career—I may not be the best son and brother, but I hope I did not let you down. Last but not least, I would like to thank my wife Mabel for her love, patience, support, and being the most invaluable person in my life.

About This book

Outdoor thermal comfort is an important consideration in the process of urban planning and urban geometry design for improving urban living quality. This book aims to introduce the assessment methods and applications of outdoor thermal comfort and contributes to the scientific understandings of urban geometry and thermal environment at neighbourhood scale using real-world examples and parametric studies. Subjective evaluation by urban dwellers and numerical modelling tools introduced in this book provide a comprehensive assessment of outdoor thermal comfort and present an integrated approach of using thermal comfort indicators as a standard in high-density cities. This book is also useful to researchers of urban climate, urban planning and design practitioners, and policymakers for more liveable urban environments.

Contents

Part III   Applications of Human Thermal Comfort in Urban Planning and Design

Chapter 1
Characteristics of Thermal Comfort in Outdoor Environments

Abstract The living quality of urban inhabitants is important to urban liveability and receives increasing concern in urban living. Thermal comfort is widely regarded as one of the important issues to urban living, particularly the health and well-being of urban inhabitants. In outdoor environments where urban dwellers spend their time for commuting, leisure, and recreational activities, the thermal environment is more complex due to the constantly changing environmental conditions and the interplay between human body and the ambient environment. Meteorological factors such as air temperature and humidity, solar radiation, and air movement are fundamental parameters of the immediate environment that one is experiencing while metabolic heat generated by human activity and clothing worn by an individual are the two personal attributes that define the human thermal environment. In outdoor environments, peoples' subjective assessment of thermal comfort is also influenced by psychological expectancy and their thermal history. The major issues associated with outdoor thermal comfort in cities include low urban wind speeds, high temperatures due to urban heat island effects, and limited solar access. In high-density cities, where complex and high-rise urban geometries are common, enhancing urban design is essential for improving outdoor thermal comfort and hence enhancing the usage of outdoor spaces.

Keywords Human thermal comfort · Outdoor environment ·
Thermophysiological · Psychological · High-density cities

1.1 Human Thermal Comfort in Outdoor Environments

Thermal comfort is important to the living quality of urban dwellers. In indoor environments such as offices and homes, thermal comfort has a wide range of benefits such as increasing productivity, reducing energy consumption, enhancing the health and well-being of building occupants. Outdoors, thermal comfort is associated with increasing use of outdoor spaces, better walkability, higher levels of physical activity, and different health outcomes. Outdoor thermal comfort therefore

© The Author(s), under exclusive license to Springer Nature Singapore Pte Ltd. 2022   1
K. K.-L. Lau et al., *Outdoor Thermal Comfort in Urban Environment*,
SpringerBriefs in Architectural Design and Technology,
https://doi.org/10.1007/978-981-16-5245-5_1

becomes an important issue in urban liveability and hence the health and well-being of urban residents.

The thermal environment which humans are exposed to is generally governed by four environmental parameters and two personal attributes. Air temperature, radiant temperature, humidity, and air movement are the four fundamental parameters of the immediate environment that one is experiencing while metabolic heat generated by human activity and clothing worn by an individual are the two personal attributes that define the human thermal environment (Parsons 1993). In indoor environments, these parameters are determined by physical settings such as building layout, orientation, materials, and occupants' behaviour. In outdoor settings, the relationship becomes more complex as it involves more frequent interactions with the surrounding environment, pedestrians or space users, and their activities (Fig. 1.1). In high-density urban environments, heavy pedestrian and vehicle traffic, as well as the complex urban geometry, result in considerable differences in thermal conditions and hence the thermoregulation of the human body.

Fig. 1.1 Different pedestrian environments in high-density cities

1.2 Factors Governing Outdoor Thermal Comfort

1.2.1 Thermophysiological Aspects of Outdoor Thermal Comfort

Six parameters are generally used to determine the thermophysiological responses of a person. Figure 1.2 describes the various components of the human body's thermoregulation. Air temperature affects the sensible heat exchange between the human body and the ambient environment while humidity affects the sweating rate and hence the evaporative cooling. Air movement increases the rate of evaporation and cooler air also increases the rate of sensible heat exchange. Radiant temperature, or mean radiant temperature as usually used in outdoor settings, is defined by two constituents, namely shortwave and longwave radiations. Shortwave radiation

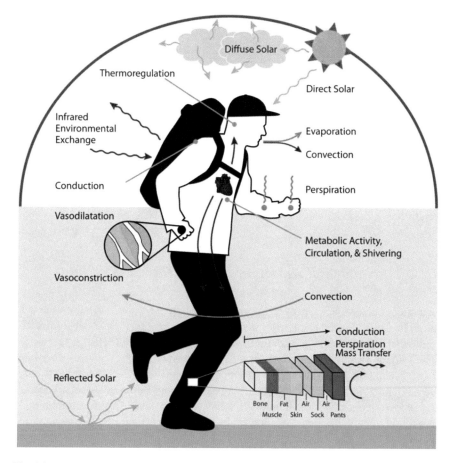

Fig. 1.2 Various components of human body's thermoregulation

includes direct solar beam radiation (or direct sun exposure), diffuse radiation from scattered incident radiation, and reflected radiation from different surfaces of built-up structures. Longwave radiation is the infrared radiation typically released from a heated object, e.g. sunlit surfaces or air-conditioning units. Personal attributes such as metabolic rate and clothing influence the mechanisms of heat exchange. Clothing plays an important role in maintaining our body temperature by serving as an insulation layer between the body and the surrounding environment. Metabolic rate, as a result of human behaviour, affects the core temperature and the associated heat transfer between body core and skin, and hence the ambient environment.

Human heat balance is based on the assumption that the heat produced within the human body over certain periods must balance the heat lost from the body. According to Nicol et al. (2012), this involves three physical processes, namely convection (heat loss through the human body warming the air around it), radiation (heat radiated to surrounding surfaces), and evaporation (heat loss through evaporating sweat and other forms of moisture). The basic thermal balance is expressed as follows:

$$M - W = C + R + E + (C_{res} + E_{res}) + S$$

M and W are the metabolic rate and the mechanical work done, respectively. C, R, and E are the convective, radiative, and evaporative heat loss from the clothed body, respectively. C_{res} and E_{res} are the convective and evaporative heat loss from respiration. S is the rate at which heat is stored in the body tissues. One of the thermophysiological heat-balance models commonly used in outdoor settings is the Munich Energy-Balance Model for Individuals (MEMI) which combines climatic parameters, metabolic activity, and type of clothing, to calculate the resultant thermal state of the human body by characterising the heat flows, body temperatures, and sweat rates. It presents a basis for the thermophysiologically relevant evaluation of the thermal bioclimate. A sample of the calculation with MEMI for warm weather conditions with direct solar irradiation is shown in Fig. 1.3.

1.2.2 Psychological Aspects of Outdoor Thermal Comfort

In outdoor environments, peoples' subjective assessment of thermal comfort is influenced by psychological expectancy and their thermal history. People felt comfortable even when they reported a "hot" thermal sensation (+3 in the ASHRAE seven-point scale) due to the abnormally cold weather in the previous days before being interviewed such that they welcomed warmer conditions (Höppe 2002). Similar findings were obtained in a study conducted on a beach (Höppe and Seidl 1991). Vacationists exposed themselves intentionally to relatively extreme thermal conditions, and it was found that the thermal sensation evaluated by heat-balance models and thermal preferences reported by the vacationists were skewed towards warmer directions.

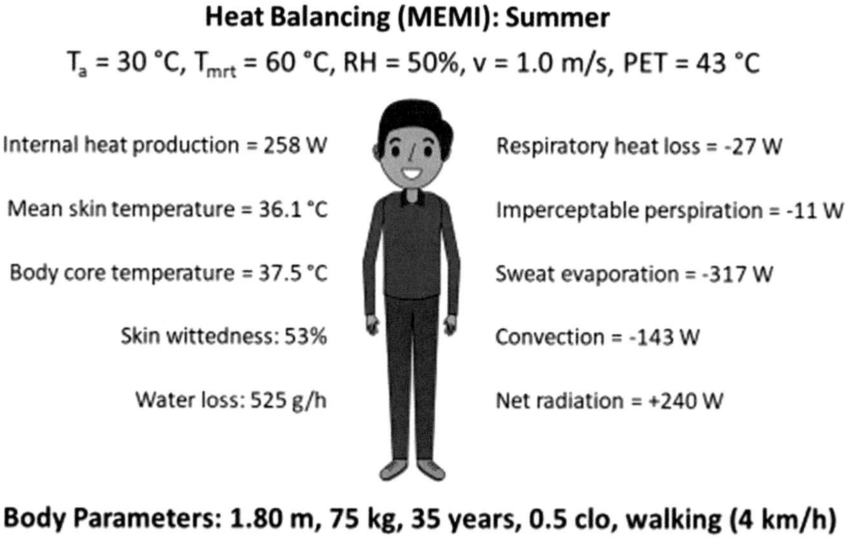

Heat Balancing (MEMI): Summer

$$T_a = 30\ °C,\ T_{mrt} = 60\ °C,\ RH = 50\%,\ v = 1.0\ m/s,\ PET = 43\ °C$$

Internal heat production = 258 W	Respiratory heat loss = -27 W
Mean skin temperature = 36.1 °C	Imperceptable perspiration = -11 W
Body core temperature = 37.5 °C	Sweat evaporation = -317 W
Skin wittedness: 53%	Convection = -143 W
Water loss: 525 g/h	Net radiation = +240 W

Body Parameters: 1.80 m, 75 kg, 35 years, 0.5 clo, walking (4 km/h)

Fig. 1.3 Sample heat-balance calculation with the MEMI model for warm and sunny condition

Such evidence shows that psychological aspects are important in subjective assessments of thermal comfort. These distinguish the thermal comfort between indoors and outdoors.

People perceive the environment in different ways according to their experience and expectation. The human response to physical stimuli is not just determined by their magnitude, but also affected by the contextual settings that people are situated in. As such, psychological factors are important to people's thermal perception of outdoor spaces and the corresponding changes in such spaces. Nikolopoulou and Steemers (2003) proposed six psychological factors that are associated with people's thermal perception in outdoor environments (Fig. 1.4). People tend to be more tolerable to changes of the physical environment that are naturally produced while their expectation of the environment has substantial influence on their thermal perception, which is defined by their past experience. On the other hand, discomfort can be tolerated if it is short-lived or if people can exert certain levels of control over the sources of discomfort. In addition, environmental stimulations are the main reason for most of the outdoor activities, and the variable nature of outdoor environments is more preferred. Nikolopoulou and Lykoudis (2006) further suggested that psychological adaptation is apparent in people's choice to respond to a source of discomfort in open spaces.

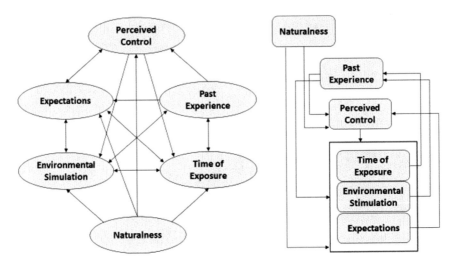

Fig. 1.4 Interrelationships (left) and the network (right) between the six psychological aspects (Nikolopoulou and Steemers 2003)

1.3 Thermal Comfort in High-Density Cities

In the urban environment, the major issues associated with thermal comfort include low urban wind speeds, high temperatures due to urban heat island effects, and limited solar access. Lower wind speed and higher air temperatures lead to thermal discomfort of people staying outdoors in the summer. Low wind speeds also hamper the ventilation potential and reduce the air flow in naturally ventilated buildings. High outdoor temperatures increase the thermal loads of buildings and cause overheating in buildings. In outdoor environments, solar radiation plays an important role in the human thermal comfort as it has opposite effects in summer and winter. In summer, exposure to solar radiation is a major source of thermal discomfort while it considerably enhances comfort in winter. In high-density cities, where complex and high-rise urban geometries are common, enhancing urban design is essential for improving outdoor thermal comfort and hence enhancing the usage of outdoor spaces.

In high-density cities, the compact form of urban development results in a wide range of environmental impacts such as poor ventilation, air pollution, and high air temperature (Ng 2009). In Hong Kong, the high-rise buildings and compact urban form lead to stagnant air in urban cores and high thermal load in the building stock. Due to the rapid and intensified urban development of the city, air temperature has increased at a rate of 0.24 °C per decade over the last 30 years, which is nearly double of the long-term rate (Fig. 1.5). Air flows in urban areas, as indicated by the wind speed observed at the King's Park meteorological station operated by the Hong Kong Observatory, have decreased from 3.5 to 2.0 m/s. Compared to the meteorological station situated on Waglan Island which does not show any significant trend, such a substantial decrease in wind speed has caused severe impacts on thermal comfort in

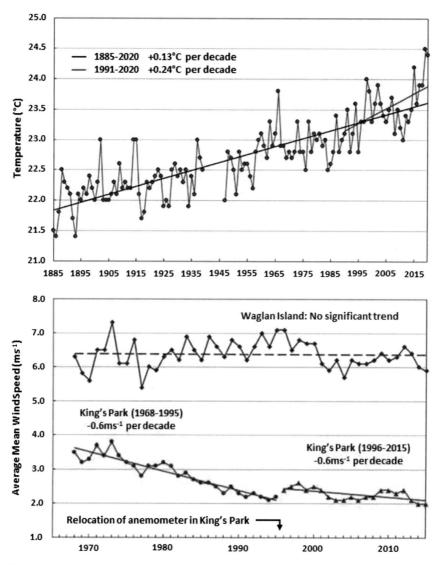

Fig. 1.5 Annual mean air temperature observed at Hong Kong Observatory Headquarter (top) and annual mean wind speed from King's Park and Waglan Island stations (bottom) (Hong Kong Observatory 2021)

urban areas. According to Ng and Cheng (2012), it is required to have wind speeds of 0.53–1.30 m/s to maintain neutral thermal conditions under shade in typical summer conditions. Considering the further decrease in wind speed at the pedestrian level, the impeded wind speed observed at King's Park station implies issues regarding the outdoor thermal comfort in the city.

1.4 Objectives of This Book

Previous studies showed that human thermal comfort in outdoor environments is significantly different from its indoor counterpart. With the effect of global climate change and rapid urbanisation, thermal conditions in urban environments are expected to be worsened, resulting in severe environmental issues that affect the comfort and well-being of urban dwellers. Therefore, knowledge of urban microclimates and the associated effect on outdoor thermal comfort is necessary for developing urban planning and design strategies that can help to improve the thermal conditions, leading to a more liveable high-density urban environment.

In order to address the issues associated with outdoor thermal comfort, subjective thermal perception of urban dwellers is required for studies to understand how people perceive the thermal environment in different weather conditions and urban settings. Questionnaire survey is the most common method, and various thermal assessment scales, such as thermal sensation (Lin et al. 2010; Krüger et al. 2017), affective evaluation of comfort (Nikolopoulou and Steemers 2003), and thermal preference (Cheung and Jim 2017), were previously used. Such subjective data are generally compared to the simultaneous meteorological conditions so that people's requirements of thermal comfort can be directly related to the observed conditions which are acquired by meteorological measurements. This contributes to the determination of thermal comfort standards in outdoor environments.

On the other hand, numerical simulations are widely used in evaluating different design scenarios and the corresponding thermal conditions of the scenarios (Liu et al. 2020). Different configurations of urban geometry can be evaluated for their effect on microclimates (Bouketta and Bouchahm 2020; Chan and Chau 2021) and their corresponding effects on thermal comfort conditions (Chatterjee et al. 2019; Ouali et al. 2020). Greening strategies are also investigated for their thermal benefits regarding buildings and surrounding neighbourhoods (Herath et al. 2018; Acero et al. 2019; Berardi et al. 2020). Numerical simulations provide an effective option for evaluating different design strategies to determine whether they meet the standard of human thermal comfort in the outdoor environment, which can be obtained by questionnaire surveys and field measurements for subjective perception of human thermal comfort.

With the increasing concern about human thermal comfort in outdoor settings, this book aims to provide a comprehensive understanding of human thermal comfort at the neighbourhood scale and to offer practical solutions for urban design strategies. In particular, practical field and numerical methods are described for the evaluation of human thermal comfort in outdoor settings, and different design strategies are introduced for mitigating outdoor heat stress and improving the thermal environment. This book is organised into three sections: human thermal comfort in the outdoor environment (Sect. 1.1); evaluation of design strategies for outdoor thermal comfort (Sect. 1.2); and practical applications of human thermal comfort (Sect. 1.3). This book eventually aims to increase the awareness of human thermal comfort in scientific, professional practice, and policymaking processes so as to create more comfortable, liveable environments for urban citizens.

References

Acero, J.A., E.J.Y. Koh, X.X. Li, L.A. Ruefenacht, G. Pignatta, and L.K. Norford. 2019. Thermal impact of the orientation and height of vertical greenery on pedestrians in a tropical area. *Building Simulation* 12 (6): 973–984.

Berardi, U., Z. Jandaghian, and J. Graham. 2020. Effects of greenery enhancements for the resilience to heat waves: A comparison of analysis performed through mesoscale (WRF) and microscale (Envi-met) modelling. *Science of the Total Environment* 747: 141300.

Bouketta, S., and Y. Bouchahm. 2020. Numerical evaluation of urban geometry's control of wind movements in outdoor spaces during winter period case of Mediterranean climate. *Renewable Energy* 146: 1062–1069.

Chan, S.Y., and C.K. Chau. 2021. On the study of the effects of microclimate and park and surrounding building configuration on thermal comfort in urban parks. *Sustainable Cities and Society* 64: 102512.

Chatterjee, S., A. Khan, A. Dinda, S. Mithun, R. Khatun, H. Akbari, H. Kusaka, C. Mitra, S.S. Bhatti, Q.V. Doan, and Y. Wang. 2019. Simulating micro-scale thermal interactions in different building environments for mitigating urban heat islands. *Science of the Total Environment* 663: 610–631.

Cheung, P.K., and C.Y. Jim. 2017. Determination and application of outdoor thermal benchmarks. *Building and Environment* 123: 333–350.

Herath, H.M.P.I.K., R.U. Halwatura, and G.Y. Jayasinghe. 2018. Evaluation of green infrastructure effects on tropical Sri Lankan urban context as an urban heat island adaptation strategy. *Urban Forestry and Urban Greening* 29: 212–222.

Hong Kong Observatory. 2021. *Climate Change in Hong Kong*. Available at: https://www.hko.gov.hk/en/climate_change/obs_hk_temp.htm

Höppe, P. 2002. Different aspects of assessing indoor and outdoor thermal comfort. *Energy and Buildings* 34 (6): 661–665.

Höppe, P., and H.A.J. Seidl. 1991. Problems in the assessment of the bioclimate for vacationists at the seaside. *International Journal of Biometeorology* 35: 107–110.

Krüger, E.L., C.A. Tamura, P. Bröde, M. Schweiker, and A. Wagner. 2017. Short- and long-term acclimatization in outdoor spaces: Exposure time, seasonal and heatwave adaptation effects. *Building and Environment* 116: 17–29.

Lin, T.P., A. Matzarakis, and R.L. Hwang. 2010. Shading effect on long-term outdoor thermal comfort. *Building and Environment* 45 (1): 213–221.

Liu, D., S. Hu, and J. Liu. 2020. Contrasting the performance capabilities of urban radiation field between three microclimate simulation tools. *Building and Environment* 175: 106789.

Ng, E. 2009. Policies and technical guidelines for urban planning of high-density cities—Air ventilation assessment (AVA) of Hong Kong. *Building and Environment* 44 (7): 1478–1488.

Ng, E., and V. Cheng. 2012. Urban human thermal comfort in hot and humid Hong Kong. *Energy and Buildings* 55: 51–65.

Nicol, F., M. Humphreys, and S. Roaf. 2012. *Adaptive Thermal Comfort: Principles and Practice*. The United Kingdom: Routledge.

Nikolopoulou, M., and S. Lykoudis. 2006. Thermal comfort in outdoor urban spaces: Analysis across different European countries. *Building and Environment* 41 (11): 1455–1470.

Nikolopoulou, M., and K. Steemers. 2003. Thermal comfort and psychological adaptation as a guide for designing urban spaces. *Energy and Buildings* 35 (1): 95–101.

Ouali, K., K. El Harrouni, M.L. Abidi, and Y. Diab. 2020. Analysis of open urban design as a tool for pedestrian thermal comfort enhancement in Moroccan climate. *Journal of Building Engineering* 28: 101042.

Parsons, K.C. 1993. *Human Thermal Environments*. The United Kingdom: Taylor and Francis.

Part I
Human Thermal Comfort in the Outdoor Environment

Chapter 2
Human Thermal Comfort in Sub-tropical Urban Environments

Abstract Outdoor thermal comfort is determined by urban morphology and the geometry of outdoor urban spaces. The local climate zone (LCZ) classification system aims to characterise the urban and rural land cover based on various urban morphological parameters. It has been widely used in studies of the thermal environment, but the subjective thermal perception between LCZ classes has rarely been studied. This study evaluated the microclimatic conditions and subjective perception of the thermal environment in eight LCZs in Hong Kong, using questionnaire surveys and field measurements. An ANOVA test showed that the microclimatic conditions were significantly different across eight LCZs, and this could be attributed to the urban morphology and the geometry of the outdoor urban spaces. This does not only affect the critical conditions but also the variations in the thermal environment. The highest maximum temperature (38.9 °C) was found in LCZ 1, and the lowest maximum temperature (29.9 °C) was observed in land cover LCZs. Subjective assessment showed that compact or high-rise settings were associated with warmer thermal sensations reported by the respondents. The relationship between the level of thermal stress and subjective thermal sensation changed across LCZs. This study demonstrated that the LCZ classification provides a characterisation of both the physical and thermal environment. It is also one of the first attempts to examine the relationship between the thermal environment and subjective perceptions using the LCZ classification system. Further work is required to investigate how thermal comfort indicators can be used to represent the thermal comfort conditions in different LCZs.

Keywords Outdoor thermal comfort · Local climate zone · Subjective thermal perception · Microclimatic conditions · Sub-tropical · High-density cities

2.1 Introduction

As a result of global climate change, air temperatures are expected to increase and subsequently causes more frequent and severe heat stress (IPCC 2014). Such intense heat stress is further exacerbated by urbanisation, leading to deteriorating levels of outdoor thermal comfort. Outdoor thermal comfort is substantially affected by the rapid urban development and the associated urban heat island (UHI) effects in sub-tropical or tropical high-density cities (Arnfield 2003). This leads to a decline in the environmental quality of outdoor urban spaces, which are generally perceived as the extended living spaces of urban citizens (Ahmed 2003).

Outdoor thermal comfort is highly influenced by urban morphology which can be quantified by the sky view factor (SVF), aspect ratio, as well as building height and coverage (Emmanuel et al. 2007; Lee et al. 2014; Lau et al. 2015). The complex nature of urban morphology has opposite effects on the thermal comfort of pedestrians as buildings provide shading and reduce ventilation (Krüger and Drach 2017). Street geometry design plays an important role in outdoor thermal comfort in tropical and sub-tropical climates (Johansson and Emmanuel 2006; Lin et al. 2013; Lau et al. 2016). Emmanuel et al. (2007), based on their field study in Sri Lanka, reported that increasing the height-to-width ratio reduces the physiological equivalent temperature (PET) by approximately 10 °C. In a numerical modelling study, Lau et al. (2016) found that the average mean radiant temperature (T_{mrt}) is considerably lower in N-S than in E-W orientated street canyons due to the prolonged exposure to solar radiation in the latter. In addition, vegetation in urban areas is an essential element to pedestrians' thermal comfort, because it provides a cooling effect through evapotranspiration and shading (Morakinyo et al. 2017).

The effect of different urban settings on pedestrians' thermal perception was first identified by Nikolopoulou and Steemers (2003). Based on questionnaire surveys, it was found that only 50% of the variation in subjective thermal sensation can be explained by climatic parameters and that site-specific context could be a key factor affecting pedestrians' thermal perception in outdoor spaces. Similar results were also reported in a study conducted in different urban settings in a sub-tropical city (Xi et al. 2012). It was found that only 40 and 30% of the variance in the thermal sensation vote (TSV) can be explained by T_{mrt} and the standard effective temperature (SET*) index, implying that elements of the built environment resulted in a large variation of the thermal environment that affects the perceived thermal comfort of pedestrians. Warmer thermal sensation was generally associated with more compact areas with more homogeneous urban morphology, as reported in a study conducted in Bangladesh (Sharmin et al. 2015). Prolonged exposure to intense solar radiation led to warmer thermal sensation than in areas with more diverse thermal environments characterised by various building height and coverage.

Krüger et al. (2017) examine the effect of the site-specific context on subjective thermal sensation. They found that the relationships between the TSV and objective thermal comfort indicators (PET and universal thermal climate index; UTCI) vary across different SVFs. This suggests that variations in urban morphology affect not only the microclimatic conditions, but also the subjective thermal perception of pedestrians. In another study, Krüger (2017) argued that such variations in subjective thermal sensation can be attributed to psychological effects, especially under moderately warm conditions. In more exposed areas under hot conditions, the variations between urban settings become minimal, suggesting that diverse urban morphologies have immense potential in mitigating thermal discomfort in urban environments. This reiterated the conceptual model proposed by Knez et al. (2009), in which the evaluation of the thermal environment includes personal experience which moderates the relationship between subjective thermal sensation and observed microclimatic conditions. Nonetheless, there is no systematic definition of different types of urban settings, such as compactness, building height, or level and types of greenery. Studies in this particular aspect may not be comparable because of their differences in the definition of outdoor urban spaces. The implications on the design of outdoor urban spaces may therefore be inconclusive.

The local climate zone (LCZ) classification system has recently been used to characterise urban and rural land cover based on a wide range of urban morphological parameters (Fig. 2.1; Stewart and Oke 2012). It is a temperature-based classification of the thermal environment based on surface characteristics including building height and coverage, pervious and impervious cover, aspect ratio, surface materials, and anthropogenic heat output (Stewart and Oke 2012). The LCZ classification system includes ten built-up classes and seven land cover types. It has been widely used in UHI studies and regional climate studies with a comprehensive parameterisation of urban surfaces (Ren et al. 2017; Skarbit et al. 2017; Cai et al. 2018). As the thermal environment of urban areas plays a key role in outdoor thermal comfort, LCZ has immense potential in studies of outdoor thermal comfort and heat stress. Previous studies used thermal comfort indicators, such as PET and UTCI, or heat indices such as HUMIDEX to quantify the differences in microclimatic conditions of LCZ classes (Ndetto and Matzarakis 2015; Geletič et al. 2018; Liu et al. 2018). Villadiego and Velay-Dabat (2014) first used the LCZ classification system to characterise the sites where an outdoor thermal comfort survey was conducted, but the differences in subjective thermal sensation across LCZ classes were not addressed. In addition, studies of thermal comfort in high-rise LCZ classes are relatively limited, as previous studies were mainly conducted in cities where such high-rise classes are uncommon (Ndetto and Matzarakis 2015; Sharmin et al. 2015).

This study aims to investigate the differences in microclimatic conditions and the subjective thermal sensation of pedestrians across LCZ classes in Hong Kong. Microclimatic conditions were obtained and compared to subjective thermal sensation data acquired by questionnaire surveys. A total of 12 survey sites covering eight LCZ classes were sampled, including four LCZ 1 and two LCZ 4 sites, as these are the most common LCZ classes in the highly urbanised areas of Hong Kong (Wang et al. 2018). The findings of the study contribute to the understanding of outdoor thermal comfort conditions under different urban contexts and address how LCZ classification systems can be used in the research of outdoor thermal comfort.

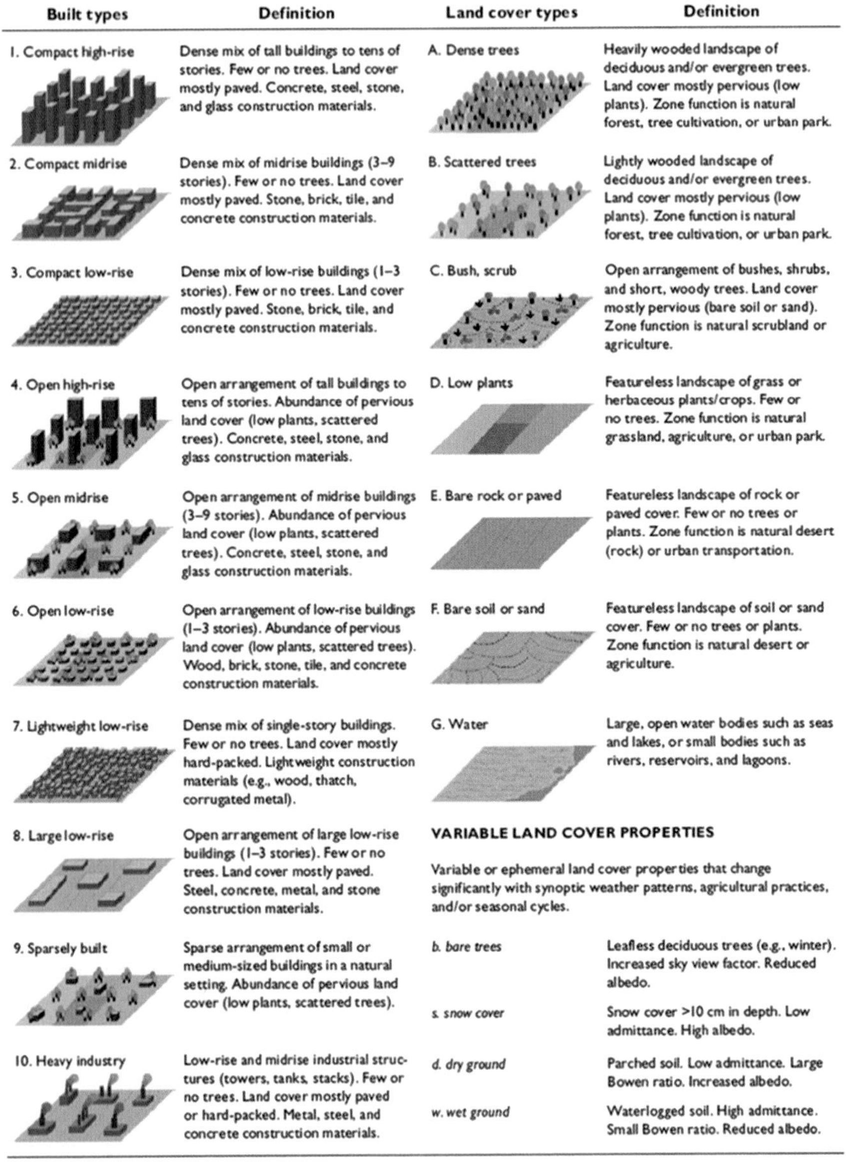

Built types	Definition	Land cover types	Definition
1. Compact high-rise	Dense mix of tall buildings to tens of stories. Few or no trees. Land cover mostly paved. Concrete, steel, stone, and glass construction materials.	A. Dense trees	Heavily wooded landscape of deciduous and/or evergreen trees. Land cover mostly pervious (low plants). Zone function is natural forest, tree cultivation, or urban park.
2. Compact midrise	Dense mix of midrise buildings (3–9 stories). Few or no trees. Land cover mostly paved. Stone, brick, tile, and concrete construction materials.	B. Scattered trees	Lightly wooded landscape of deciduous and/or evergreen trees. Land cover mostly pervious (low plants). Zone function is natural forest, tree cultivation, or urban park.
3. Compact low-rise	Dense mix of low-rise buildings (1–3 stories). Few or no trees. Land cover mostly paved. Stone, brick, tile, and concrete construction materials.	C. Bush, scrub	Open arrangement of bushes, shrubs, and short, woody trees. Land cover mostly pervious (bare soil or sand). Zone function is natural scrubland or agriculture.
4. Open high-rise	Open arrangement of tall buildings to tens of stories. Abundance of pervious land cover (low plants, scattered trees). Concrete, steel, stone, and glass construction materials.	D. Low plants	Featureless landscape of grass or herbaceous plants/crops. Few or no trees. Zone function is natural grassland, agriculture, or urban park.
5. Open midrise	Open arrangement of midrise buildings (3–9 stories). Abundance of pervious land cover (low plants, scattered trees). Concrete, steel, stone, and glass construction materials.	E. Bare rock or paved	Featureless landscape of rock or paved cover. Few or no trees or plants. Zone function is natural desert (rock) or urban transportation.
6. Open low-rise	Open arrangement of low-rise buildings (1–3 stories). Abundance of pervious land cover (low plants, scattered trees). Wood, brick, stone, tile, and concrete construction materials.	F. Bare soil or sand	Featureless landscape of soil or sand cover. Few or no trees or plants. Zone function is natural desert or agriculture.
7. Lightweight low-rise	Dense mix of single-story buildings. Few or no trees. Land cover mostly hard-packed. Lightweight construction materials (e.g., wood, thatch, corrugated metal).	G. Water	Large, open water bodies such as seas and lakes, or small bodies such as rivers, reservoirs, and lagoons.
8. Large low-rise	Open arrangement of large low-rise buildings (1–3 stories). Few or no trees. Land cover mostly paved. Steel, concrete, metal, and stone construction materials.	**VARIABLE LAND COVER PROPERTIES**	
		Variable or ephemeral land cover properties that change significantly with synoptic weather patterns, agricultural practices, and/or seasonal cycles.	
9. Sparsely built	Sparse arrangement of small or medium-sized buildings in a natural setting. Abundance of pervious land cover (low plants, scattered trees).	b. bare trees	Leafless deciduous trees (e.g., winter). Increased sky view factor. Reduced albedo.
		s. snow cover	Snow cover >10 cm in depth. Low admittance. High albedo.
10. Heavy industry	Low-rise and midrise industrial structures (towers, tanks, stacks). Few or no trees. Land cover mostly paved or hard-packed. Metal, steel, and concrete construction materials.	d. dry ground	Parched soil. Low admittance. Large Bowen ratio. Increased albedo.
		w. wet ground	Waterlogged soil. High admittance. Small Bowen ratio. Reduced albedo.

Fig. 2.1 Definition for local climate zones (Stewart and Oke 2012)

2.2 Methodology

2.2.1 The Climate and LCZ Map of Hong Kong

Hong Kong is located at 22° 15′ N 114° 10′ E with a sub-tropical monsoon climate. It is typically hot and humid in the summer with a summer mean air temperature of 28 °C. Air temperatures in the afternoon often exceed 31 °C while relative humidity ranges from 60 to 70% during the daytime in the summer. There is more than 50% of possible sunshine from July to December, with daily mean global solar radiation peaking near 200 W m^{-2} in July (Hong Kong Observatory 2015). Hong Kong is also famous for being a high-density city. Its density can be interpreted in two aspects, namely population and urban density. Hong Kong had a population of 7.4 million in 2017, with a population density of approximately 6,700 persons per km^2 in its urban area (Census and Statistics Department 2018). The other aspect is its high-density and high-rise urban morphology in built-up areas, with an average building height of 60 m (Ng et al. 2011). The compact urban form results in intense UHI effect and insufficient air ventilation.

The LCZ scheme (Stewart and Oke 2012) has become a worldwide standard in UHI studies to classify urban morphologies and nature landscapes. Recently, Hong Kong researchers have explored different methods to develop LCZ classifications and maps in Hong Kong (Siu and Hart 2013; Wang et al. 2018; Zheng et al. 2018). Since this study focuses on the neighbourhood scale, the LCZ map of Wang et al. (2018) with a finer spatial resolution of 100 m and higher accuracy has been adopted here. Hong Kong has a diverse urban landscape in which 17 standard LCZs were identified (Fig. 2.2). A large proportion of land areas are country parks classified as the vegetated LCZ A to D, and high-rise LCZ classes (LCZ 1 and 4) are dominant in the city centre. LCZ 1–6, D (low plants), F (bare soil or sand) were selected to conduct micrometeorological measurements and an outdoor thermal comfort survey in the present study (Table 2.1). Other LCZ classes were not assessed, owing either to the absence of regular pedestrians or the lack of accessible clusters in these areas.

2.2.2 Micrometeorological Measurements

Micrometeorological measurements were conducted between 10 and 16 h on selected days from June to September 2018 to obtain the conditions which the survey respondents were exposed to. The questionnaire survey and micrometeorological measurements were carried out at designated sites on clear summer days with similar weather conditions to minimise the variations across survey days. Two mobile meteorological stations were used at the same time (Fig. 2.3), with each containing a TESTO 480 data logger for measuring air temperature (T_a), relative humidity (RH), and wind speed (v) and a globe thermometer for measuring the globe temperature (T_g). The globe thermometer is composed of a thermocouple wire (TESTO flexible Teflon type

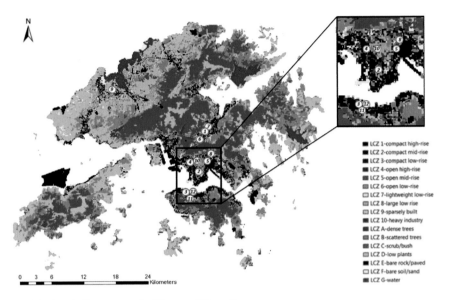

Fig. 2.2 Local climate zone (LCZ) map of Hong Kong

K) held inside a black table tennis ball with a diameter (D) of 38 mm and emissivity (ε) of 0.95. Mean radiant temperature (T_{mrt}) is subsequently determined using the following equation from Thorsson et al. (2007):

$$T_{mrt} = \left[\left(T_g + 273.15 \right)^4 + \frac{1.10 \times 10^8 * v^{0.6}}{\varepsilon * D^{0.4}} \left(T_g - T_a \right) \right]^{1/4} - 273.15 \qquad (2.1)$$

In the present study, PET was used as the thermal index to evaluate the thermal conditions that respondents experienced. It is based on the Munich Energy-Balance Model for Individuals and defined as the "air temperature at which the heat balance of the human body is maintained with core and skin temperature equal to those under the conditions being assessed" (Höppe 1999). PET has been widely used in thermal comfort studies in different climatic regions and may allow for interurban comparisons in future studies (Coccolo et al. 2016). It is calculated by considering T_a, RH, v, and T_{mrt}, as well as the participant's metabolic rate and clothing level recorded by the questionnaire survey.

2.2.3 Questionnaire Surveys

A thermal comfort questionnaire survey was carried out simultaneously with microm-eteorological measurements at designated sites in order to obtain the subjective thermal sensation of respondents. The survey was conducted from June to September

Table 2.1 LCZ classes, aerial and fish-eye photos, and major characteristics of the survey sites

LCZ classes	Aerial and fish-eye photos	Major characteristics of the survey sites
LCZ 1 Compact High-rise		• Commercial or mixed-use districts • High level of pedestrian and vehicle traffic • Lack of vegetation and limited open spaces
LCZ 2 Compact mid-rise		• Residential areas with ground-level retail shops • Several high-rise buildings were inaugurated in the last five to ten years • Moderate level of pedestrian and vehicle traffic • A large urban park in the northern part of the area
LCZ 3 Compact low-rise		• Low-rise residential buildings or individual houses • Low level of pedestrian and vehicle traffic • Lower population in the area
LCZ 4 Open high-rise		• Public housing estates • More abundant vegetation and open spaces • High level of pedestrian traffic during rush hour • Low level of vehicle traffic • Leisure activities commonly found in open spaces
LCZ 5 Open mid-rise		• Mixed-use district • Lack of vegetation • Moderate level of pedestrian and vehicle traffic
LCZ 6 Open low-rise		• Low-rise private housing • Low-density residential area • Low level of pedestrian and vehicle traffic
LCZ D Low plants		• Large urban parks located on the hillslope • Abundant vegetation • Leisure activities are common

(continued)

Table 2.1 (continued)

LCZ classes	Aerial and fish-eye photos	Major characteristics of the survey sites
LCZ F Bare soil or sand		• High-level of pedestrian traffic near the piers • Limited vegetation but large extent of open spaces along the waterfront

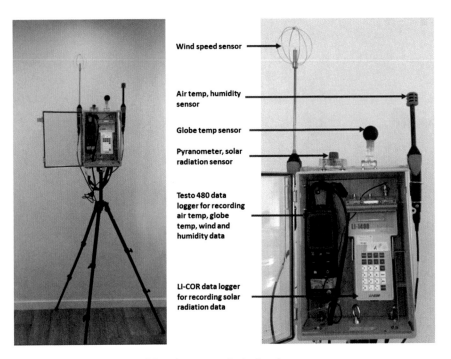

Fig. 2.3 Instrumental set-up of the micrometeorological station

2018, and it took place between 10 and 16 h on weekdays to minimise variations in weather conditions and pedestrian activities. A total of 1917 effective questionnaires were obtained and used in the subsequent statistical analysis. The questionnaire consisted of questions on sensation regarding temperature (TSV), humidity (HSV), wind speed (WSV), and solar radiation (SSV) based on the seven-point ASHRAE scale (ASHRAE 2010), such that thermal sensations were reported from cold (−3) to hot (+3), with neutral sensation as 0. Humidity sensations were reported from very dry (−3) to very wet (+3), and wind and solar sensations were reported from very weak (−3) to very strong (+3). The overall state of thermal comfort was rated on a four-point Likert scale from very uncomfortable (−2) to very comfortable (+2), without any option for the neutral state. The demographic background of the participants was also obtained, and the activity level of the participants was recorded to

Table 2.2 Characteristics of survey respondents

Sex	n	%	15-min AC environment	n	%
Male	863	45.0	Yes	682	35.6
Female	1054	55.0	No	1235	64.4
Age	**n**	**%**	**15-min activity**	**n**	**%**
<18	480	25.0	Sitting	237	12.4
18–24	363	18.9	Standing	637	33.2
25–34	212	11.1	Walking	1026	53.5
35–44	175	9.1	Doing exercise	17	0.9
45–54	186	9.7			
>55	491	25.6	Total no. of respondents	1917	
Prefer not revealed	10	0.5			

represent the metabolic rate. The clothing levels were observed by the interviewers using the checklists from ANSI/ASHRAE Standard 55 (ASHRAE 2010).

The respondent characteristics are presented in Table 2.2. The numbers of male and female respondents are 863 (45.0%) and 1054 (55.0%), respectively, and young (<18) and old (>55) respondents account for approximately half of the respondents. 35.6% of the respondents were in air-conditioned environment 15 min before the survey. Most of the respondents (53.5%) were walking in the last 15 min, and respondents who were sitting and standing 15 min prior to the survey account for 12.4% and 33.2%, respectively. As the survey was conducted during weekdays and mostly in street environments, only 0.9% of the respondents were doing exercise before conducting the survey.

2.2.4 Statistical Analysis

An analysis of variance (ANOVA) was performed to determine whether the means of meteorological variables and corresponding subjective perception were significantly different among LCZs. A post hoc Tukey's test was conducted to identify pairs of LCZ which were significantly different in PET and/or subjective evaluation of the thermal environment. Linear regression was then used to examine the correlation between mean TSV and meteorological variables (Krüger et al. 2015; Lam and Lau 2018).

2.3   Results and Discussion

2.3.1   Micrometeorological Conditions of Different LCZs

The meteorological variables measured in the different LCZs are shown in Table 2.3. With the similar weather conditions of the survey days, it provided a meaningful comparison of survey sites with different urban morphological settings. The mean T_a of the built-up LCZs (LCZ 1–6) measured during the survey days was slightly higher than that of the land cover LCZs (LCZ D and F). The largest range of T_a was observed in LCZ 1 (8.46 °C), due to the high variations in urban morphology within the highly built-up areas. The generally higher buildings in this LCZ class resulted a large amount of heat stored in the building mass. However, the high variation in T_a is predominantly due to the highly variable building height in the survey locations. The smallest range was found in LCZ 2 (2.90 °C) since the measurements were taken in a compact but relatively homogeneous environment dominated by buildings with five to six storeys. The relatively lower SVF (0.37) is another reason for the homogeneous T_a in this class. The highest maximum T_a was also measured in LCZ 1, emphasising

Table 2.3 Measurements of meteorological variables in different LCZs

	LCZ 1	LCZ 2	LCZ 3	LCZ 4	LCZ 5	LCZ 6	LCZ D	LCZ F
Air temperature (°C)								
Maximum	38.9	35.1	36.9	35.7	37.3	37.2	35.8	35.4
Mean	33.7	33.5	34.0	32.6	34.9	34.0	31.4	32.2
Minimum	30.4	32.2	32.5	29.8	32.2	31.8	29.9	31.3
Relative humidity (%)								
Maximum	77.8	64.3	61.6	89.3	73.6	64.5	80.8	74.0
Mean	63.8	57.4	55.6	67.9	60.0	57.7	73.0	67.4
Minimum	45.6	48.6	50.2	53.6	52.9	49.2	55.6	51.4
Wind speed (ms^{-1})								
Maximum	2.4	3.2	1.6	2.2	2.2	1.0	1.4	2.5
Mean	1.0	1.6	0.8	1.0	1.0	0.6	0.5	1.1
Minimum	0.1	0.2	0.4	0.1	0.4	0.3	0.1	0.4
T_{mrt}(°C)								
Maximum	73.7	81.2	66.9	64.6	69.3	56.4	53.8	64.5
Mean	40.1	55.8	40.6	35.1	42.1	37.0	34.3	36.4
Minimum	30.4	37.5	32.5	29.8	32.2	31.8	29.9	31.3
PET (°C)								
Maximum	56.4	55.9	52.3	51.2	53.4	47.9	44.6	49.7
Mean	37.2	44.0	37.7	33.8	39.2	36.0	33.0	33.8
Minimum	29.4	36.5	32.1	29.1	32.4	31.7	29.8	30.5

the high thermal load due to extensive urban structures and anthropogenic activities such as pedestrian and vehicle traffic, street food stalls, and waste heat from air conditioning systems. The maximum T_a measured in LCZ D and F were lower (35.8 and 35.4, respectively) because of the cooling effect of extensive urban vegetation and the close proximity to the waterfront. Lower minimum T_a were measured in high-rise LCZs (LCZ 1 and 4) as well as in the two land cover LCZs (LCZ D and F), owing to the shading effect of either buildings or vegetation, which reduced the exposure to intense solar heat.

The mean of RH of different LCZs ranged from 55.6 to 73.0% with higher maximum RH observed in LCZ 4 and D owing to the presence of vegetation in the survey sites. The highest minimum RH was also found in the densely vegetated LCZ D. Lower minimum RH was observed in compact urban settings. High mean v was measured in LCZ F because of the close proximity to the waterfront and the influence of sea breezes. In LCZ 2, v was generally higher owing to the presence of a large urban park near the survey location. In particular, the minimum v was found to be higher in LCZ 3 and 6, because of the lower building coverage in the proximity of the survey locations, resulting in higher wind permeability and ventilation potential.

Larger ranges of T_{mrt} were observed in compact urban settings (43.28 °C and 43.71 °C for LCZ 1 and 2, respectively). The highest maximum T_{mrt} was also observed in these two LCZs, suggesting that the complex urban structures created large variations in urban morphology and hence solar exposure. Meanwhile, the mean T_{mrt} of LCZ 1 was considerably lower than that of LCZ 2 by 27.8%, owing to the higher shading opportunity by surrounding high-rise buildings. The considerably lower SVF in LCZ 1 (0.15) also accounted for the difference. More than 40% of the samples in LCZ 2 showed T_{mrt} higher than 55 °C and this could lead to heat stress due to intense solar exposure (Thorsson et al. 2014). Similar maximum T_{mrt} values were also found in the exposed waterfront (LCZ F). The lowest T_{mrt} was observed in LCZ D, primarily owing to the presence of dense vegetation cover which reduced shortwave radiation through shading and longwave radiation owing to shaded surfaces.

PET was generally influenced by the level of T_{mrt} in most of the LCZ classes. LCZ 2 exhibited the highest mean and minimum PET (44.0 °C and 36.5 °C, respectively) because of the more exposed environment. In a compact urban environment, the high proportion of impervious and building surfaces absorbs a large amount of heat and increases the ambient temperature and longwave radiation from these warm surfaces, leading to less favourable thermal comfort conditions. The largest range of PET was measured in LCZ 1, whereas the smallest range was found in LCZ D. Lower PET values were measured in LCZ D and F, owing to the presence of vegetation and the proximity to the waterfront. Meanwhile, thermal comfort conditions were more variable in open LCZs, with the mean PET of LCZ 4 (33.8 °C) being considerably lower than those of the other two open LCZs (39.2 °C and 36.0 °C). Mid-rise LCZs were the most uncomfortable areas, with the influence of high T_a and less adequate shading opportunities by surrounding buildings.

The ANOVA test results (Table 2.4) showed that the means of all the meteorological variables (i.e. T_a, RH, v, T_{mrt}) and PET were significantly different among

Table 2.4 Results of analysis of variance (ANOVA) for different meteorological variables

Variable		Sum of squares	df	Mean square	F	P-value
T_a	Between groups	1343.5	7	191.9	120.4	<0.0001
	Within groups	3048.9	1913	1.6		
	Total	4392.4	1920			
RH	Between groups	46,076.4	7	6582.3	187.1	<0.0001
	Within groups	67,305.7	1913	35.2		
	Total	113,382.2	1920			
v	Between groups	132.8	7	19.0	176.6	<0.0001
	Within groups	205.7	1915	0.1		
	Total	338.5	1922			
T_{mrt}	Between groups	61,643.7	7	8806.2	129.3	<0.0001
	Within groups	130,324.1	1913	68.1		
	Total	191,967.9	1920			
PET	Between groups	16,749.2	7	2392.7	112.8	<0.0001
	Within groups	40,565.8	1913	21.2		
	Total	57,315.0	1920			

the LCZs, suggesting that the micrometeorological conditions of the LCZs were significantly different. This demonstrates that the difference in urban morphology resulted in a wide variety of meteorological conditions such as shading, air ventilation, and exposure to shortwave and longwave radiation. A post hoc Tukey's test was conducted to determine the significantly different pairs of LCZs. Most of the T_a pairs were significantly different, which demonstrates the strength of LCZ classification in urban temperature or UHI studies (Stewart and Oke 2012). However, the differences among compact LCZs (LCZ 1–3) were not significant. This is likely due to the complex urban environment in Hong Kong, characterised by large podium structures, high-rise buildings, and a compact but varying urban morphology (Fig. 2.4). Nonetheless, most of the LCZ pairs for RH and v were found to be significantly different, except for a few pairs with LCZ 2, owing to the presence of a large urban park near the survey location. For T_{mrt}, the differences between LCZ 4, 6, D, and F were not significant, owing to the relatively open urban morphology in LCZ 4 and 6, which resulted in a high level of exposure to solar radiation; this was also observed in LCZ D and F (large urban parks and waterfront, respectively). Four notable insignificant pairs of LCZs were observed in PET. LCZ 3 was found to be insignificantly different from LCZ 5 and 6, and LCZ 4 were insignificantly different from D and F. This can be attributed to the more open spaces available in LCZ 3 and 4 (urban low-rise areas and public housing estates), and it reiterates the dominance of solar radiation in the urban thermal environment.

Fig. 2.4 Typical characteristics of urban morphology in Hong Kong: **a** podium structure at ground level; **b** high-rise and compact form; **c** large variations in building height, and **d** lack of vegetation

2.3.2 Subjective Perception of Microclimatic Conditions in Different LCZs

Information about the subjective perception of microclimatic conditions was obtained in eight LCZs of Hong Kong, with the aim to understand how these conditions were subjectively assessed by pedestrians. Figure 2.5 shows the distribution of the subjective perception votes of the four meteorological parameters, namely temperature (TSV), humidity (HSV), wind speed (WSV), and solar radiation (SSV).

The majority of TSVs fell into the "hot" (+3) and "warm" (+2) categories (Fig. 2.5a). The largest proportion of these two categories was found in LCZ 1 and 4 (77.7% and 76.2%, respectively). Such results were expected in LCZ 1 because of the large amount of heat trapped in the compact environment, consistent with the field measurements. T_a was generally lower in LCZ 4, but the proportion of high TSVs was larger. This was likely due to the longer exposure to high temperature in public housing estates, where the respondents spent a prolonged period in the outdoor spaces. They tended to stay in the spaces for longer period (e.g. 1–2 h), and the prolonged exposure to high temperature resulted in higher thermal sensation, which was also observed in other studies conducted in urban spaces (Lin 2009; Krüger and Drach 2017). In LCZ D and F, there was a smaller proportion of high TSVs observed (59.5% and 64.2%, respectively) where extensive vegetation and close proximity to the waterfront lead to cooler sensation, even under hot summer conditions. In contrast, respondents reported the highest proportion of "slightly warm" (+1), "neutral" (0), and "slightly cool" (−1) votes in LCZ D, 2, and F (40.5%, 36.9%, and

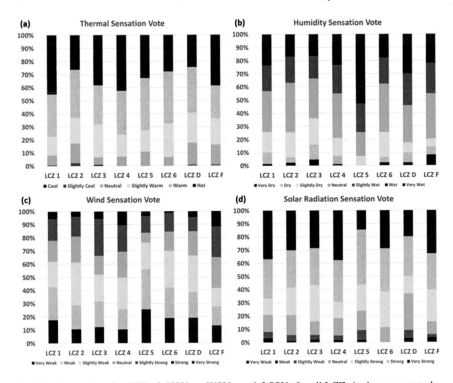

Fig. 2.5 Proportion of **a** TSVs, **b** HSVs, **c** WSVs, and **d** SSVs for all LCZs in the present study

35.8%, respectively). Apart from the two land cover LCZs, LCZ 2 exhibited more neutral thermal conditions among the built LCZs. This could be due to the influence of a large urban park near the survey site, which improves respondents' subjective thermal perception. It also had the largest proportion of neutral TSV among all the LCZs. In terms of overall thermal comfort, the two land cover LCZs had nearly 50% of the respondents reporting comfort votes, and LCZ 1 and 5 had the least proportion of respondents perceiving the environment as comfortable.

TSVs were averaged for every 0.5 °C in T_a, and the data were binned separately for compact, open and land cover LCZs in order to show the effects of LCZ classes on the TSV reported by the respondents. Figure 2.6 shows the linear trends of the relationship between the mean TSV and T_a (Fig. 2.6a), and PET (Fig. 2.6b) for the three LCZ groups. For T_a, the slope of the regression line of the compact LCZs was higher than that of the open LCZs, suggesting that the respondents were more sensitive to the changes in T_a under more compact urban settings. In the range of lower T_a, the mean TSVs of the open LCZs were higher than those of the compact LCZs, owing to the influence of the more exposed environment. An inverse trend was observed in higher-T_a range, which could be attributed to lower air speed in the compact urban areas, resulting in warmer TSVs reported by the respondents. The intersection point can be observed at approximately 34.5 °C in the regression lines

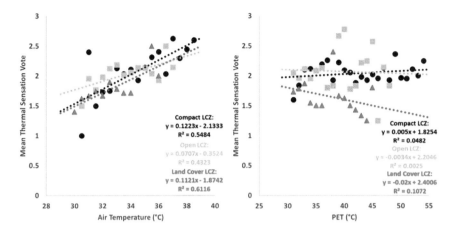

Fig. 2.6 Binned TSV for **a** every 0.5 °C of Ta and **b** every 1 °C of PET

of the mean TSV and T_a for compact and open LCZ groups. Kruger et al. (2017) observed this difference in the relationship between the mean TSV and PET. They argued that beyond a certain threshold, which is defined by the intersection point of the two regression lines for high and low SVF, the relationship between the level of thermal stress and the corresponding thermal sensation is reversed. However, in the present study, the relationship between the mean TSV and PET is not significant, particularly for the open LCZs. This is because the respondents were exposed to both shaded and sunlit conditions during the questionnaire survey.

Humidity is associated with the thermoregulatory mechanism of human body because high humidity inhibits the evaporation of sweat and reduces the sweating rate. Although the observed RH was relatively high during the survey days (over 60% on average), respondents reporting a neutral HSV accounted for approximately 30% of the total respondents across all the LCZs, except for LCZ 5 with a proportion of 18.2% observed (Fig. 2.5b). If the "neutral" range were expanded to include "slightly dry" and "slightly wet", it would cover approximately 70% of the built-up LCZs, except LCZ 5, as well as 60.3% and 63.5% for LCZ D and F, respectively. The highest proportion of the respondents perceiving the environment as "very dry" (-3) or "dry" (-2) was observed in LCZ 3 and LCZ F (16.3% and 14.6%, respectively). The sensation "very wet" ($+3$) or "wet" ($+2$) had approximately 20% of the total votes across all the LCZs, except for LCZ 5 (52.7%). The high proportion of wet sensation in LCZ 5 is likely due to the study area being located on a relatively flat area surrounded by higher mountains and close to the shore. It also has a river channel next to the survey location, leading to the wet sensation reported by the respondents.

The HSV was compared to the TSV in order to examine the possible relationship between the sensations of temperature and humidity. 13.4% of the neutral HSV were found in the neutral TSV group. The majority of the neutral HSV was found in warmer TSV groups, with 33.2% and 30.5% in "warm" and "hot" votes, respectively. 33.0% of the total respondents reported wet HSVs and warm TSVs at the same time. This suggests that the inverse relationship between air temperature and relative humidity is not applicable to the subjective perception of pedestrians. Villadiego and Velay-Dabat (2014) argued that the role that humidity plays in respondents' perception is not well understood by themselves, and they attribute their perception to temperature without considering the effect of relative humidity. However, the respondents of the present study are not well adapted to a humid climate, unlike those in Makaremi et al. (2012) and Villadiego and Velay-Dabat (2014). This could be attributed to the pre-existing perception of summer heat and humidity in Hong Kong, leading to independent assessments of heat and humidity.

Air ventilation is important for outdoor thermal comfort in sub-tropical climates because the high solar angle during summer results in nearly complete sun exposure in street canyons and compact urban areas (Ng et al. 2011). The heat accumulated within the urban environment can only be removed through air movement. As shown in Fig. 2.5c, the majority of the respondents reported weak wind sensations across all the eight LCZs, with the highest proportion of weak WSVs observed in LCZ 5 (76.4%). This is the reason for the high proportion of wet HSVs in the same LCZ, because people tend to feel that stagnant air leads to the accumulation of moist air. 70% and 66.4% of the weak WSVs were observed in LCZ 6 and D, respectively, but relatively fewer "warm" and "hot" TSVs were observed in these two LCZs. This suggests that the thermal and wind sensations do not have a direct relationship, which was further confirmed in LCZ 3 and 4 where the respondents reported higher proportions of strong WSVs but also high "warm" and "hot" TSVs. Interestingly, neutral WSVs accounted for the least proportion in all LCZs, suggesting that the respondents are relatively sensitive to air movement.

Solar radiation plays a key role in human thermal comfort in outdoor environments (Höppe 2002). The radiative environment is highly dependent on urban geometry, which affects direct and indirect (e.g. reflected) radiation (Lau et al. 2015). In the present study, approximately 80% of the respondents reported either "slightly strong", "strong", or "very strong" SSVs in all eight LCZs, except for LCZ D where only 63% of the respondents felt that the solar radiation was strong (Fig. 2.5d). The dense vegetation in the park provided a more sheltered environment, and thus they experienced intermittent sun exposure. The highest proportion of strong SSVs was found in LCZ 6, where no weak votes were reported. This LCZ is dominated by low-rise individual houses without many shelters on the street. The intense solar exposure may exaggerate people's perception to solar radiation, as they do not have sufficient relief or shelter from sun exposure. The TSVs were cross tabulated with the SSVs to examine the relationship between thermal and solar radiation sensations. It was found that thermal sensation is highly associated with solar radiation sensation. 24.3% of the total respondents reported both "hot" TSV and "very strong" SSV, and a further 16.4% of the total respondents reported both "warm" TSV and "strong" SSV. In contrast, only 4.2% of the total respondents reported both neutral TSV and SSV, showing the dominance of intense heat during summer afternoons in Hong Kong.

2.3.3 Correlation Between Subjective Perception and Meteorological Variables

A correlation analysis was conducted to examine the relationship between subjective perception of micrometeorological conditions and the measured values of meteorological variables (Table 2.5). Among the measured values of meteorological variables, all the correlations are significant at $\alpha = 0.01$ level. Unsurprisingly, T_a and RH had a strong and inverse relationship ($r^2 = -0.830$, p-value < 0.0001). There was a weak relationship between T_a and v ($r^2 = 0.067$, p-value < 0.0001) but v was moderately correlated with T_{mrt} ($r^2 = 0.475$, p-value < 0.0001), which was likely due to the more open settings resulting in higher sun exposure and air movement. T_a was also moderately correlated with T_{mrt} ($r^2 = 0.401$, p-value < 0.0001) and T_{mrt} was highly correlated with PET ($r^2 = 0.973$, p-value < 0.0001), reiterating the influence of solar radiation on the level of thermal comfort.

There was a weak but statistically significant correlation between TSV and T_a ($r^2 = 0.126$, p-value < 0.0001), but TSV was not significantly correlated with RH. SSV was significantly correlated with all meteorological variables, despite the weak correlations. On the contrary, the correlations between the subjective perception and measured meteorological variables were all significant, although the correlations were relatively weak (temperature: $r = 0.126$; humidity: $r = 0.118$; wind: $r = 0.123$; radiation: $r = 0.087$). The findings of the present study are consistent with previous studies, in that individual meteorological parameters cannot explain perceptions of the thermal environment (Höppe 2002; Nikolopoulou and Lykoudis 2006). Psychological factors may play a key role in the thermal perception of outdoor spaces in different urban settings, especially in high-density cities.

2.4 Conclusions

The present study evaluated the microclimatic conditions and subjective perception of the thermal environment in eight LCZs in Hong Kong, using questionnaire surveys and field measurements conducted during summer 2017. A total of 1917 responses to effective questionnaires consisting of questions on the subjective assessment of sensations regarding temperature, humidity, wind speed, and solar radiation were obtained. An ANOVA test showed that the microclimatic conditions were significantly different across the eight LCZs, and this can be attributed to variations in the urban morphology and geometry of the outdoor spaces. It not only affects the critical conditions but also the variations in the thermal environment. The highest maximum T_a (38.9 °C) was found in LCZ 1, while the lowest maximum T_a (29.9 °C) was observed in land cover LCZs. The largest range of T_a was observed in LCZ 1 (8.46 °C), owing to the great variations in urban morphology within the highly built-up areas.

Table 2.5 Correlation matrix between subjective perception and measured values of meteorological variables

	TSV	SSV	WSV	HSV	T_a	RH	v	T_{mrt}	PET
TSV	/								
SSV	0.534	/							
WSV	-0.131	-0.080	/						
HSV	0.031	-0.021	-0.015	/					
T_a	0.126	0.184	-0.066	-0.087	/				
RH	-0.045	-0.126	0.012	0.118	-0.830	/			
V	0.055	0.124	0.123	-0.074	0.067	-0.224	/		
T_{mrt}	0.031	0.087	0.006	-0.103	0.475	-0.446	0.401	/	
PET	0.039	0.090	-0.012	-0.098	0.376	-0.567	0.586	0.973	/

Note that Pearson correlations were used for the correlations among meteorological variables while Spearman correlations were used for the correlations among subjective perception and between subjective perception and measured values of meteorological variables

A subjective assessment of the thermal environment showed that compact or high-rise settings were associated with warmer thermal sensations reported by the respondents. As indicated by the PET, mid-rise LCZs had the most undesirable thermal comfort conditions owing to inadequate shading opportunities in outdoor spaces. Vegetation and proximity to the waterfront also influence thermal sensations, demonstrating the cooling effect present in the land cover LCZs. It was found that the relationship between the level of thermal stress and subjective thermal sensation would change under different urban settings. The subjective perception of micro-climatic conditions was significantly correlated with the corresponding measured meteorological variables. Significant correlations were found between TSV and T_a ($r = 0.126$), and SSV was correlated with all the meteorological variables. The correlations between subjective perception and measured meteorological variables were all significant, although the correlations are relatively weak (temperature: $r = 0.126$; humidity: $r = 0.118$; wind: $r = 0.123$; radiation: $r = 0.087$). The weak relationship suggests that physical factors such as meteorological conditions cannot fully explain the perception of thermal environment, and psychological factors may play a key role.

This study demonstrated that the LCZ classification system provides a characterisation of both the physical and thermal environments. It was also one of the first attempts to examine the relationship between the thermal environment and the corresponding subjective perceptions, based on the LCZ classification system. Limitations of the study include the limited samples in some of the LCZ classes. The findings could be further enhanced by including more study sites in the future. As the study focused on the summer conditions, which are the most critical to outdoor thermal comfort in Hong Kong, it would be useful to ascertain how winter conditions affect outdoor thermal comfort in different LCZ classes in Hong Kong. Further work is required to investigate how thermal comfort indicators such as PET can be used to represent the thermal comfort conditions in different LCZs, as the present study did not find a useful relationship between PET and TSV.

References

Ahmed, K.S. 2003. Comfort in urban spaces: Defining the boundaries of outdoor thermal comfort for the tropical urban environments. *Energy and Buildings* 35 (1): 103–110.

Arnfield, A.J. 2003. Two decades of urban climate research: A review of turbulence, exchanges of energy and water, and the urban heat island. *International Journal of Climatology* 23: 1–26.

ASHRAE. 2010. *ANSI/ASHRAE Standard 55-2010. Thermal Environmental Conditions for Human Occupancy*. Atlanta: American Society of Heating, Refrigerating and Air-Conditioning Engineers, Inc.

Cai, M., C. Ren, Y. Xu, K.K.L. Lau, and R. Wang. 2018. Investigating the relationship between local climate zone and land surface temperature using improved a WUDAPT methodology. *Urban Climate* 24: 485–502.

Census and Statistics Department. 2018. *Annual Digest of Statistics 2017*. Hong Kong: Census and Statistics Department.

Coccolo, S., J. Kämpf, J.L. Scartezzini, and D. Pearlmutter. 2016. Outdoor human comfort and thermal stress: A comprehensive review on models and standards. *Urban Climate* 18: 33–57.

Emmanuel, R., H. Rosenlund, and E. Johansson. 2007. Urban shading—A design option for the tropics? A study in Colombo, Sri Lanka. *International Journal of Climatology* 27: 1995–2004.

Geletič, J., M. Lehnert, S. Savić, and D. Milošević. 2018. Modelled spatiotemporal variability of outdoor thermal comfort in local climate zones of the city of Brno, Czech Republic. *Science of the Total Environment* 624: 385–395.

Hong Kong Observatory. 2015. Monthly Meteorological Normals for Hong Kong. Available at: https://www.hko.gov.hk/cis/normal/1981_2010/normals_e.htm#table6.

Höppe, P. 1999. The physiological equivalent temperature—A universal index for the biometeorological assessment of the thermal environment. *International Journal of Biometeorology* 43 (2): 71–75.

Höppe, P. 2002. Different aspects of assessing indoor and outdoor thermal comfort. *Energy and Buildings* 34 (6): 661–665.

IPCC. 2014. *Climate Change 2014: Synthesis Report. Contribution of Working Groups I, II and III to the Fifth Assessment Report of the Intergovernmental Panel on Climate Change* [Core Writing Team, R.K. Pachauri and L.A. Meyer (eds.)]. Geneva, Switzerland: IPCC.

Johansson, E., and R. Emmanuel. 2006. The influence of urban design on outdoor thermal comfort in the hot, humid city of Colombo, Sri Lanka. *International Journal of Biometeorology* 51 (2): 119–133.

Knez, I., S. Thorsson, I. Eliasson, and F. Lindberg. 2009. Psychological mechanisms in outdoor place and weather assessment: Towards a conceptual model. *International Journal of Biometeorology* 53 (1): 101–111.

Krüger, E. 2017. Impact of site-specific morphology on outdoor thermal perception: A case-study in a subtropical location. *Urban Climate* 21: 123–135.

Krüger, E.L., and P. Drach. 2017. Identifying potential effects from anthropometric variables on outdoor thermal comfort. *Building and Environment* 117: 230–237.

Krüger, E., P. Drach, and P. Bröde. 2015. Implications of air-conditioning use on thermal perception in open spaces: A field study in downtown Rio de Janeiro. *Building and Environment* 94: 417–425.

Krüger, E., P. Drach, and P. Bröde. 2017. Outdoor comfort study in Rio de Janeiro: Site-related context effects on reported thermal sensation. *International Journal of Biometeorology* 61 (3): 463–475.

Lam, C.K.C., and K.K.L. Lau. 2018. Effect of long-term acclimatization on summer thermal comfort in outdoor spaces: A comparative study between Melbourne and Hong Kong. *International Journal of Biometeorology* 62 (7): 1311–1324.

Lau, K.K.L., F. Lindberg, D. Rayner, and S. Thorsson. 2015. The effect of urban geometry on mean radiant temperature under future climate change: A study of three European cities. *International Journal of Biometeorology* 59 (7): 799–814.

Lau, K.K.L., C. Ren, J. Ho, and E. Ng. 2016. Numerical modelling of mean radiant temperature in high-density sub-tropical urban environment. *Energy and Buildings* 114: 80–86.

Lee, H., H. Mayer, and D. Schindler. 2014. Importance of 3-D radiant flux densities for outdoor human thermal comfort on clear-sky summer days in Freiburg, Southwest Germany. *Meteorologische Zeitschrift* 23 (3): 315–330.

Lin, T.P. 2009. Thermal perception, adaptation and attendance in a public square in hot and humid regions. *Building and Environment* 44 (10): 2017–2026.

Lin, T.P., K.T. Tsai, C.C. Liao, and Y.C. Huang. 2013. Effects of thermal comfort and adaptation on park attendance regarding different shading levels and activity types. *Building and Environment* 59: 599–611.

Liu, L., Y. Lin, Y. Xiao, P. Xue, L. Shi, X. Chen, and J. Liu. 2018. Quantitative effects of urban spatial characteristics on outdoor thermal comfort based on the LCZ scheme. *Building and Environment* 143: 443–460.

Makaremi, N., E. Salleh, M.Z. Jaafar, and A. GhaffarianHoseini. 2012. Thermal comfort conditions of shaded outdoor spaces in hot and humid climate of Malaysia. *Building and Environment* 48: 7–14.

Morakinyo, T.E., L. Kong, K.K.L. Lau, C. Yuan, and E. Ng. 2017. A study on the impact of shadow-cast and tree species on in-canyon and neighborhood's thermal comfort. *Building and Environment* 115: 1–17.

Ndetto, E.L., and A. Matzarakis. 2015. Urban atmospheric environment and human biometeorological studies in Dar es Salaam, Tanzania. *Air Quality, Atmosphere & Health* 8 (2): 175–191.

Ng, E., C. Yuan, L. Chen, C. Ren, and J.C.H. Fung. 2011. Improving the wind environment in high-density cities by understanding urban morphology and surface roughness: A study in Hong Kong. *Landscape and Urban Planning* 101 (1): 59–74.

Nikolopoulou, M., and S. Lykoudis. 2006. Thermal comfort in outdoor urban spaces: Analysis across different European countries. *Building and Environment* 41 (11): 1455–1470.

Nikolopoulou, M., and K. Steemers. 2003. Thermal comfort and psychological adaptation as a guide for designing urban spaces. *Energy and Buildings* 35 (1): 95–101.

Ren, C., J.C.H. Fung, J.W.P. Tse, R. Wang, M.N.F. Wong, and Y. Xu. 2017. Implementing WUDAPT product into urban development impact analysis by using WRF simulation result—A case study of the Pearl River Delta Region (1980–2010). In *Proceedings of the 13th Symposium on Urban Environment*, Seattle, WA.

Sharmin, T., K. Steemers, and A. Matzarakis. 2015. Analysis of microclimatic diversity and outdoor thermal comfort perceptions in the tropical megacity Dhaka, Bangladesh. *Building and Environment* 94: 734–750.

Siu, L.W., and M.A. Hart. 2013. Quantifying urban heat island intensity in Hong Kong SAR, China. *Environmental Monitoring and Assessment* 185 (5): 4883–4893.

Skarbit, N., I.D. Stewart, J. Unger, and T. Gál. 2017. Employing an urban meteorological network to monitor air temperature conditions in the 'local climate zones' of Szeged, Hungary. *International Journal of Climatology* 37: 582–596.

Stewart, I.D., and T.R. Oke. 2012. Local climate zones for urban temperature studies. *Bulletin of the American Meteorological Society* 93 (12): 1879–1900.

Thorsson, S., F. Lindberg, I. Eliasson, and B. Holmer. 2007. Different methods for estimating the mean radiant temperature in an outdoor urban setting. *International Journal of Climatology* 27: 1983–1993.

Thorsson, S., J. Rocklöv, J. Konarska, F. Lindberg, B. Holmer, B. Dousset, and D. Rayner. 2014. Mean radiant temperature—A predictor of heat related mortality. *Urban Climate* 10: 332–345.

Villadiego, K., and M.A. Velay-Dabat. 2014. Outdoor thermal comfort in a hot and humid climate of Colombia: A field study in Barranquilla. *Building and Environment* 75: 142–152.

Wang, R., C. Ren, Y. Xu, K.K.L. Lau, and Y. Shi. 2018. Mapping the local climate zones of urban areas by GIS-based and WUDAPT methods: A case study of Hong Kong. *Urban Climate* 24: 567–576.

Xi, T., Q. Li, A. Mochida, and Q. Meng. 2012. Study on the outdoor thermal environment and thermal comfort around campus clusters in subtropical urban areas. *Building and Environment* 52: 162–170.

Zheng, Y., C. Ren, Y. Xu, R. Wang, J. Ho, K. Lau, and E. Ng. 2018. GIS-based mapping of local climate zone in the high-density city of Hong Kong. *Urban Climate* 24: 419–448.

Chapter 3
Dynamic Response of Pedestrian Thermal Comfort

Abstract Outdoor thermal comfort studies have proved that urban design has a great influence on pedestrians' thermal comfort and that its assessment helps one to understand the quality and usage of the pedestrian environment. However, the majority of outdoor thermal comfort studies perceive pedestrian thermal comfort as "static". The dynamic multiple uses of urban spaces and the highly inhomogeneous urban morphology in high-density cities of the tropics are seldom considered, which leads to a lack of understanding about how pedestrians respond to the changes of the outdoor environment. This study contributes to the understanding of the dynamic thermal comfort using a longitudinal survey that was conducted to obtain information about how thermal sensation changes throughout the walking route and how it is affected by micrometeorological conditions and the urban geometry. The large variations in micrometeorological conditions throughout the walking routes are predominantly influenced by the urban geometry. Additionally, the spatial pattern of thermal sensation varies based on the weather conditions, emphasising the need to account for such variations in the assessment of pedestrian thermal comfort. The results also show that thermal sensation was associated with participants' short-term thermal experience (2–3 min) and that the urban geometry plays an important role in the time-lag effect of meteorological variables on thermal sensation. The findings of this study contribute to improving urban geometry design in order to mitigate the thermal discomfort and create a better pedestrian environment in high-density cities.

Keywords Outdoor thermal comfort · Transient · Pedestrian environment · high-density cities

3.1 Introduction

Urban liveability and the health and well-being of urban citizens are highly associated with the outdoor thermal environment, especially in the era of pedestrianisation which promotes a healthier lifestyle and a more sustainable urban environment (Castillo-Manzano et al. 2014). As a result of complex urban geometry, there are large variations of microclimatic conditions in urban areas, which in turn affects the

K. K.-L. Lau et al., *Outdoor Thermal Comfort in Urban Environment*,
SpringerBriefs in Architectural Design and Technology,
https://doi.org/10.1007/978-981-16-5245-5_3

level of thermal comfort experienced by pedestrians and hence their activities and behaviours (Krüger et al. 2011). Evaluation of outdoor thermal comfort is therefore essential to improve the quality of the outdoor thermal environment and, more importantly, the use of outdoor spaces (Maruani and Amit-Cohen 2007).

The heat balance of the human body and its heat exchange with the surrounding environment is the basis of human thermal comfort. A wide range of numerical indices such as predicted mean vote (PMV; Fanger 1973), physiological equivalent temperature (PET; Höppe 1999), and Universal Thermal Climate Index (UTCI; Bröde et al. 2012) were developed to incorporate environmental and personal parameters including air temperature, air humidity, air velocity, mean radiant temperature (T_{mrt}), clothing insulation, and level of activity, into the prediction of thermal sensation (Taleghani et al. 2015). Such indices are developed on the basis of the exposure of the human body to steady environmental conditions that are not common in outdoor environments (Höppe 2002). As such, the dynamic nature of outdoor environmental conditions must be incorporated into the assessment of human thermal comfort in the outdoor environment.

The changes of thermal sensation in response to a transient environment were previously studied under indoor conditions. Gagge et al. (1967) first discovered the overshooting in thermal sensation occurred during a human body moving from a warm environment to a neutral environment and the lags occurred in the opposite direction. Subsequent studies also reported this phenomenon (de Dear et al. 1993; Nagano et al. 2005; Chen et al. 2011). The magnitude of the step change was also proved to affect thermal responses (Xiong et al. 2015; Yu et al. 2015) and subjective thermal sensations often stabilise faster than physiological responses like skin temperature (Arens et al. 2006; Chen et al. 2011). It was also suggested that thermal transients result in different responses of the thermal regulation of human body (Potvin 2000). The rate of transition is important to maintaining or improving human thermal comfort under constantly changing environmental conditions.

Outdoor thermal comfort studies conventionally focused on how meteorological conditions affect instantaneous subjective thermal sensation (Spagnolo and de Dear 2003; Pantavou et al. 2013). However, pedestrians are often exposed to constantly changing environmental conditions due to complex urban geometries in the outdoor environment. Using the Instationary Munich Energy-Balance Model (IMEM), Höppe (2002) explained the differences in human thermophysiology between outdoor and indoor thermal comfort. Based on the model simulation using a "sunny street segment" scenario that a pedestrian leaving a shaded area of a sidewalk and entering a sunny segment of 200 m length, he argued that steady-state thermal comfort models are only applicable for persons in an outdoor environment for more than 30 min because skin temperature gradually increases when a person enters an outdoor environment and approaches the skin temperature predicted by the steady-state model after 180 s (Fig. 3.1). It is therefore possible to avoid prolonged hot thermal sensation if the urban geometry is carefully designed without exceeding this threshold.

In a subsequent study, similar human thermal responses were also obtained when the subjects left an indoor environment and entered different outdoor scenarios (Katavoutas et al. 2015). A walking study was conducted in European cities to

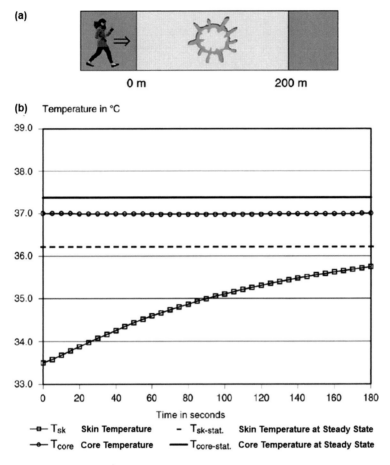

Fig. 3.1 **a** "Sunny street segment" scenario used for the IMEM model simulation; **b** temporal variation of the skin (T_{sk}) and core (T_{core}) temperature calculated from the IMEM model after entering a sunny segment on a hot summer day. The horizontal lines represent T_{sk} and T_{core} at steady-state levels (Höppe 2002)

investigate the changes in pedestrian thermal comfort in the outdoor environment (Vasilikou and Nikolopoulou 2013). It found that pedestrians could perceive the variations in the environmental conditions during the walking routes. In Japan, a mobile measurement system was previously employed to record the micrometeorological conditions and individuals' physiological responses along a predefined pedestrian route covering a wide range of urban geometries and surface environments (Nakayoshi et al. 2015). Thermal sensation was found to be influenced by the cutaneous thermoreceptors responding to subtle environmental changes (de Dear 2011), reiterating the importance of pedestrians' physiological responses and thermal histories.

In this study, the dynamic response of pedestrians' thermal sensations between two designated routes in a high-density commercial area of Hong Kong was investigated, taking into account the effect of urban geometry and corresponding micrometeorological conditions. A longitudinal survey was conducted to obtain subjective thermal sensation and how it changed throughout the walking routes, as well as how urban geometry and micrometeorological conditions affect the changes in thermal sensation. Findings of this study are important for the knowledge of urban geometry design based on how pedestrians may tolerate the discomfort during their walk. Urban designers can consider the dynamic responses of pedestrian thermal comfort when they design outdoor spaces to enhance the walking environment in high-density cities.

3.2 Methodology

3.2.1 Longitudinal Survey

A longitudinal survey was conducted to obtain information about the dynamic response of pedestrians travelling in the urban environment and how it is associated with urban geometry and the corresponding micrometeorological conditions. The survey campaigns were carried out in a high-density commercial area of Hong Kong and consisted of two walking routes with the same starting and destination points to cover the variations of the urban geometry (Fig. 3.2a). Route 1 generally consisted of narrow street canyons with occasionally open spots at street intersections and open

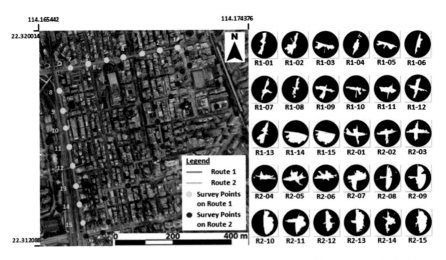

Fig. 3.2 **a** Two walking routes and 15 surveys points along each route of the present study; **b** fish-eye photos of each survey point

Table 3.1 Meteorological conditions of the days when the survey being conducted

Date	Time	Air temp (°C)	Rel. hum. (%)	Wind direction	Mean wind speed (m/s)	Sky condition	Mean Amt of cloud (%)
8-Aug-16	14:22–15:22	31.6–32.8	69–73	NW	3.6	Overcast sky	83
22-Aug-16	14:21–15:10	31.3–31.8	64–72	SE	3.1	Clear sky	27
13-Sep-16	13:54–14:54	29.5–30.5	70–76	SE	2.8	Partially cloudy	61
14-Sep-16	13:53–14:39	31.7–32.8	50–60	N, W	1.9	Partially cloudy	59

spaces, while Route 2 basically followed the two main roads along the E-W orientation and then the N-S orientation. The duration of the walking routes was approximately 60 min, and the total length was approximately 1.5 km. The participants instantaneous started the walking routes in order to minimise the temporal differences of the background meteorological conditions. There were 15 survey points in each walking route. The participants were asked for their thermal sensation vote (TSV) for the environment that they were exposed to, based on the ASHRAE seven-point scale from cold (-3) to hot ($+3$). Sky view factor (SVF) was used to represent the compactness of the urban geometry, and the SVF values were calculated for each survey point using the RayMan model (Matzarakis and Rutz 2010) based on fish-eye photos taken during the survey (Fig. 3.2b).

The survey campaigns were conducted on four late summer days to capture three types of weather conditions (clear, partially cloudy, and overcast sky). They were carried out from 2 h to 4 h to represent the critical summer conditions in Hong Kong. Participants included six male and eight female university students with ages from 19 to 21. Table 3.1 details the meteorological conditions of the surveys, which were obtained from the ground-level meteorological stations operated by the Hong Kong Observatory and located less than 1 km away from the study area.

3.2.2 Mobile Meteorological Measurements

The mobile meteorological station used in the study was installed in a backpack for measuring meteorological conditions during the survey campaigns (Fig. 3.3). Microclimatic sensors were installed in the backpack for measuring air temperature (T_a), relative humidity (RH), and wind speed (v). RH was then converted into absolute humidity (AH) using Eq. (3.1):

$$AH = \frac{6.112 \times e^{\left[\frac{17.67 \times T}{T+243.5}\right]} \times RH \times 2.1674}{273.15 + T} \quad (3.1)$$

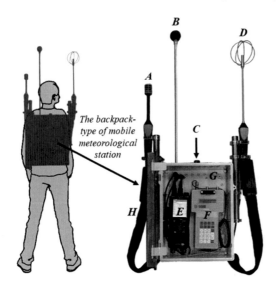

A. **Air temperature and relative humidity sensor (protected from direct sunlight)**
B. **Globe thermometer**
C. **Pyranometer**
D. **Wind speed sensors**
E. **Data logger for air temperature, relative humidity and wind speed sensors, and globe thermometer**
F. **Data logger for pyranometer**
G. **Protection case**
H. **Straps**

Fig. 3.3 Instrumental set-up of the mobile meteorological station used in the present study

A tailor-made globe thermometer was used to measure globe temperature (T_g), which is composed of a thermocouple wire held in the middle of a 38-mm black table tennis ball in order to minimise the response time during mobile measurements (Humphreys 1977; Nikolopoulou et al. 1999). Table 3.2 shows the technical specifications of the instruments used in the present study such as their corresponding

Table 3.2 Sensors used to measure meteorological parameters in the present study

Parameter	Sensor	Measurement range	Accuracy	Measured interval
T_a	Air temperature and relative humidity sensors	−20 to +70 °C	±0.2 °C (+15 °C to +30 °C) ± 0.5 °C (remaining range)	1 s
RH	Air temperature and relative humidity sensors	0–100% RH	±(1.0% + 0.7% of measured value) (0 to 90% RH) ±(1.4–0.7% of measured value) (90–100% RH) ±0.03% RH/K (based on 25 °C)	1 s
v	Wind speed sensor	0 to +5 m/s	±(0.03 m/s + 4% of measured value)	1 s
T_g	Globe thermometer	−50 to +250 °C	±1.1 °C-or 0.4% of reading)	1 s

T_a—air temperature; RH—relative humidity; v—wind speed; T_g—globe temperature

range and parameters. Mean radiant temperature (T_{mrt}) was estimated using Eq. (3.2) (Thorsson et al. 2007):

$$T_{mrt} = \left[\left(T_g + 273.15 \right)^4 + \frac{1.10 \times 10^8 * v^{0.6}}{\varepsilon * D^{0.4}} \left(T_g - T_a \right) \right]^{1/4} - 273.15 \qquad (3.2)$$

where ε is the emissivity (0.95 for a black globe) and D is the globe diameter. The backpack was carried by a student helper to avoid additional load on the participants.

PET was calculated from the above meteorological and human parameters and was used as an objective indicator of pedestrian thermal comfort. It is based on the "Munich Energy-balance Model for Individuals (MEMI), which models the thermal conditions of the human body in a physiologically relevant way" (Höppe 1999; p. 71). PET is defined as the air temperature that "in a typical indoor setting, the heat balance of the human body is maintained with core and skin temperatures equal to those under the conditions being assessed." (Höppe 1999; p. 73). Participants were asked to wear typical summer clothing to maintain a clothing value below 0.5 clo, and no observable change in clothing was recorded during the survey. The level of metabolic activities was assumed to be 2.0 met, which represents a slow walking speed of 2.0 km/h for pedestrian activities in a commercial/shopping district (Fanger 1973).

3.2.3 Statistical Analysis

The possible time-lag effect of thermal sensation (i.e. the immediate thermal experience of participants) was investigated by using autocorrelation analysis. Average TSVs obtained at survey points were used as the inputs of the autocorrelation analysis represented by Eq. (3.3).

$$r_k = \frac{\sum_{i=1}^{N-k} \left(X_i - \overline{X} \right) \left(X_{i+k} - \overline{X} \right)}{\sum_{i=1}^{N} \left(X_i - \overline{X} \right)^2} \qquad (3.3)$$

r is the autocorrelation which measures the linear dependency among the process variables X_i and X_{i+k}, and k represents the number of lags concerned in the analysis. N represents the total number of survey points. The partial autocorrelation at lag k is also defined as the direct correlation between X_i and X_{i+k} with the linear dependence between the intermediate variables removed. The entire time series was considered in the autocorrelation analysis, and significant correlations would be determined in order to examine whether immediate thermal history plays a role in instantaneous thermal sensation.

Additionally, TSV was used as the dependent variable for the cross-correlations with meteorological variables, such as T_a, AH, v, and T_{mrt}, as well as thermal comfort indicator, PET, in order to determine the time lag(s) of the effect of the meteorological variables preceding the thermal sensation reported by subjects. The cross-correlation function between the two sequences x and y (r_{xy}) is represented by Eq. (3.4).

$$r_{xy} = \frac{1}{L\sigma_x\sigma_y} \sum_{t=0}^{L-1} x(t)y(t+h) \tag{3.4}$$

σ_x and σ_y are the corresponding standard deviation of the two sequences, and L is the length of sequence. 95% of confidence interval was employed in the present study to assess the significance of autocorrelations and cross-correlations.

3.3 Results and Discussion

3.3.1 Micrometeorological Measurements

Considerable spatiotemporal variations in meteorological conditions were found along the walking routes during the survey. Figure 3.4 shows the spatiotemporal variations in T_a, AH, v, T_{mrt} and the PET on 13 September 2016 (a partly cloudy day). The influence of the urban morphology and street geometry was reiterated by the highly fluctuating T_a and T_{mrt} measured along the walking routes. T_a was found to be higher in more open sections of the walking routes, for example, a higher T_a (>32 °C) was observed from the 10th to the 22th minute while travelling Route 1 when the subjects passed through a long section of a N-S orientated street (Fig. 3.5a). Direct exposure to intense solar radiation due to the high sun altitude resulted in the high level of T_{mrt}.

High levels of pedestrian activities also contributed to the high T_a measured along the walking routes which can be characterised by high pedestrian and vehicle traffic, retail activities including hawkers and street food stalls, which result in high level of anthropogenic heat release. The latter half of Route 1 mostly covered narrow streets, which provided sufficient shading (Fig. 3.5b). A lower T_a was found in this section except at the 45th minute when the subject reached an open intersection. Nonetheless, there were no large variations in the absolute humidity due to the relatively lower variation in activities that lead to considerable changes of the moisture content of air.

Highly variable wind speed (v) was measured throughout the walking route due to the complex urban morphology of the study area. It was relatively higher (>1 m/s) in the section next to a large sports ground (12th to 14th minute; Fig. 3.5c) and the intersection with a main road (30th to 34th minute; Fig. 3.5d) along Route 1. Interestingly, v was relatively higher in the E-W narrow streets in the latter half of Route 1 (up to 1.5 m/s) possibly due to the closer proximity to a waterfront and an alignment with the prevailing wind direction in the study area. In a compact urban

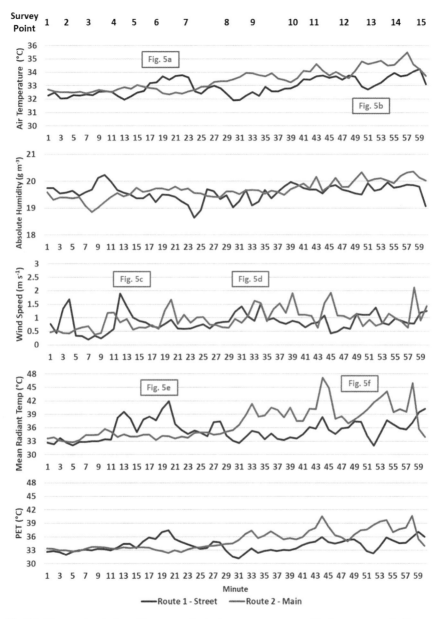

Fig. 3.4 Temporal variation of the meteorological parameters measured in the present study. Survey points and the location of the photos in Fig. 3.5 were also denoted

Fig. 3.5 **a** A long section of a N-S orientated street along Route 1; **b** shading provided by buildings in the latter half of Route 1; **c** exposed environment due to the presence of a large sports ground; **d** open street intersection in the middle section of Route 1; **e** shaded side of the first half of Route 2; and **f** tree shades along the latter half of Route 2

environment, urban planning and building design are important to the permeability of urban fabrics, which highly influence air movement (Yuan and Ng 2012).

In the first half of Route 2, lower T_a and T_{mrt} were measured because the subjects were walking along the shaded side of an E-W orientated main road (Fig. 3.5e). When the subjects turned onto a N-S orientated road, T_a and T_{mrt} increased to 33.8 °C and 36.2 °C, respectively. However, roadside trees occasionally provided shading to this particular section of the route (Fig. 3.5f), which considerably decreased T_{mrt} by approximately 5 °C. Additionally, the higher wind speed measured in the latter half of

Route 2 also corresponded to the lower air temperature, emphasising the importance of shading and ventilation in the microclimate of street environments in high-density cities.

3.3.2 Spatial Variation of the PET and Subjective Thermal Sensation

The spatial variation of the PET along the two walking routes under different skies was shown in Fig. 3.6. In general, high values of TSVs were observed along the more exposed main road and areas close to large open spaces, except for during overcast conditions for which the variations of the TSV and PET diminished. Under partially cloudy conditions, PET values were up to 36 °C along the main road (Route 2), where the subjects were largely exposed. The corresponding TSV that was reported by the subjects was up to +2.5, indicating a thermally uncomfortable environment. Locations with trees present showed lower PET values of approximately 28 °C and served as a "break" for urban dwellers to recover from the heat stress they had experienced. The PET values were consistently low in the narrow streets (Route 1) due to the shading by surrounding high-rise buildings. The participants reported mostly neutral thermal sensation throughout the entire route.

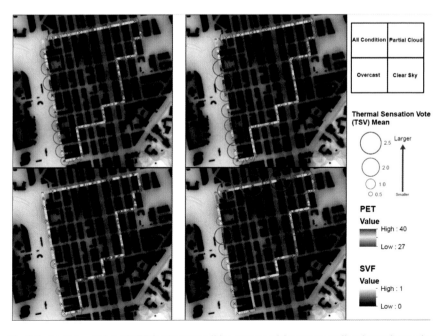

Fig. 3.6 Spatial variation of PET for the two walking routes and the corresponding thermal sensation votes reported by the respondents

Consistently high PET values were observed along Route 2 under clear sky conditions, and most of the participants voted for warm (+2) to hot (+3) for thermal sensation. However, there were some lower TSVs reported between the higher TSV values. The immediate thermal history of the participants was the reason for such a large contrast, resulting in the much lower TSVs that were reported by the participants. This phenomenon was previously suggested as "thermal alliesthesia" (Parkinson and de Dear 2015), which implies that there are certain thresholds that pedestrians may be able to tolerate. This is important to the design of urban geometries since it offers greater flexibility to the design without compromising the thermal comfort of pedestrians.

3.3.3 Autocorrelation of Subjective Thermal Sensation

Nikolopoulou and Steemers (2003) argued that the immediate short-term thermal experience affects a person's thermal sensation due to the conditions of the surrounding environment, which is particularly relevant to the design of an urban geometry. The short-term thermal experience can range from minutes to hours and influence the thermal sensation of a new environment through involuntary comparison with the previous experience (Ji et al. 2017). Figure 3.7 shows that insignificant autocorrelations were found for both walking routes as the autocorrelations were within the 95% confidence interval (i.e. null hypothesis is not rejected). However, the lag-1 autocorrelation (0.355, p-value $= 0.129$) was marginally insignificant for the trajectory of thermal sensation obtained from Route 2, which indicates that there is a possible effect of the thermal sensation of the previous survey point (2-3 min before) on the instant one reported by the subjects. Such an effect was also demonstrated by the high correlation observed in the partial autocorrelation function, indicating that there is potentially a direct, positive correlation between the instantaneous and preceding thermal sensations. This implies a possible level of tolerance to changing environmental conditions that can be incorporated into the design of an urban geometry and is particularly useful to areas with environmental constraints. It also suggests that future studies can further look into the effect of immediate thermal history on people's thermal sensation in the outdoor environment.

3.3.4 Time-Lag Effect of Meteorological Variables on the TSV

In order to examine the time-lag effect of meteorological variables on the TSV throughout the walking route under all weather conditions, cross-correlation analysis was conducted to determine the correlation between thermal sensation and different lags of the meteorological variables (Fig. 3.8). For Route 1, there were no significant

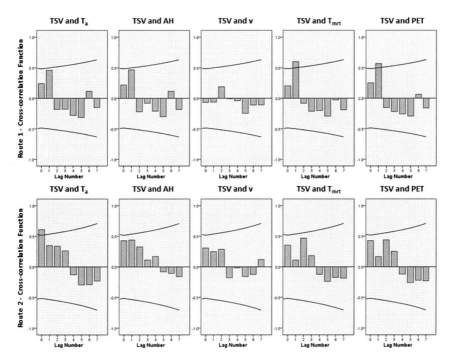

Fig. 3.7 Autocorrelation function (ACF) and partial ACF of thermal sensation votes reported by the respondents for the two walking routes. The solid lines denote the 95% confidence interval which determines whether the null hypothesis is rejected

correlations between the TSV and the instantaneous values of the meteorological variables. However, significant correlations were found between the TSV and the lag-1 values of T_{mrt} and the PET ($r = 0.600$ and 0.567, respectively). The correlations between the TSV and the lag-1 T_a and AH were marginally insignificant, suggesting that meteorological conditions may induce a delayed response to human thermal sensation when pedestrians are in motion and that the "memory" of recent conditions affects satisfaction with the thermal environment (Nikolopoulou et al. 2001). This short-term "memory" of the thermal experience was further confirmed by the insignificant correlations between the TSV and the lag-2 values (and thereafter) of the meteorological variables.

For Route 2, a significant correlation was observed between the TSV and the instantaneous value of T_a ($r = 0.608$), while the cross-correlations with the instant T_{mrt} and the PET are less significant. In contrast, the cross-correlations were insignificant between the TSV and the lag-1 values of T_{mrt} and the PET. A more homogeneous and exposed environment may reduce the influence of short-term changes of the environmental conditions since the pedestrians were continuously exposed to solar radiation in the latter half of the route, which is a wider road exposed to the sun in the afternoon. This fact suggests that, in a continuously exposed environment, thermal

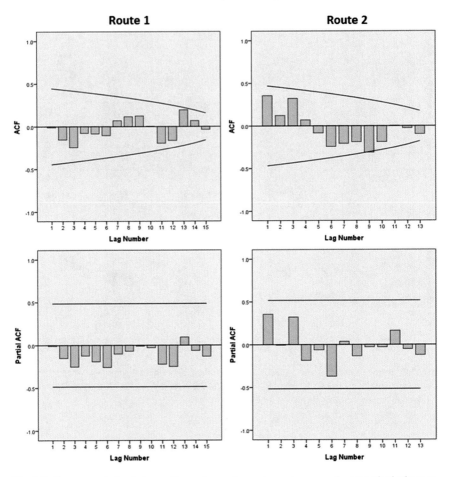

Fig. 3.8 Cross-correlation between thermal sensation votes and measured meteorological parameters. The solid lines denote the 95% confidence interval which determines whether the null hypothesis is rejected

sensation may be overwhelmed by instantaneous environmental conditions, while a constantly changing environment may have a higher potential to allow pedestrians to seek favourable conditions during their walk.

3.4   Conclusions

The dynamic nature of outdoor thermal comfort was investigated by using a longitudinal survey and field measurements of micrometeorological conditions obtained by a mobile measurement system. Survey campaigns consisting of two designated walking routes were conducted in a dense urban area of Hong Kong. The results show that there are considerable variations in the meteorological conditions and the corresponding thermal sensation reported by the subjects. Openness is one of the predominant factors influencing pedestrians' thermal comfort. The subjects also felt more comfortable when moving from sunlit to shaded places. The improvement of thermal comfort was greater when the difference in the PET was higher. The spatial pattern of thermal sensation was also found to vary with the weather conditions, demonstrating the need account for such variations in the assessment of pedestrian thermal comfort. The results also show that thermal sensation was associated with participants' short-term thermal experience (2–3 min) and that urban geometry plays an important role in the time-lag effect of meteorological variables on thermal sensation. This implies that more careful geometry design is important to pedestrian-level thermal comfort, accounting for the level of tolerance to thermal discomfort. Further data collection will be conducted to refine the understanding of such time-lag effects of tolerance and the physiological mechanism behind them.

References

Arens, E., H. Zhang, and C. Huizenga. 2006. Partial-and whole-body thermal sensation and comfort—Part I: Uniform environmental conditions. *Journal of Thermal Biology* 31 (1): 53–59.

Castillo-Manzano, J.I., L. Lopez-Valpuesta, and J.P. Asencio-Flores. 2014. Extending pedestrianization processes outside the old city center: Conflict and benefits in the case of the city of Seville. *Habitat International* 44: 194–201.

Chen, C.P., R.L. Hwang, S.Y. Chang, and Y.T. Lu. 2011. Effect of temperature steps on human shin physiology and thermal sensation response. *Building and Environment* 46: 2387–2397.

de Dear, R. 2011. Revisiting an old hypothesis of human thermal perception: Alliesthesia. *Building Research and Information* 39 (2): 108–117.

de Dear, R.J., J.W. Ring, and P.O. Fanger. 1993. Thermal sensations resulting from sudden ambient temperature changes. *Indoor Air* 3: 181–192.

Fanger, P.O. 1973. Assessment of man's thermal comfort in practice. *British Journal of Industrial Medicine* 30 (4): 313–324.

Gagge, A.P., J.A.J. Stolwijk, and J.D. Hardy. 1967. Comfort and thermal sensations and associated physiological responses at various ambient temperatures. *Environmental Research* 1: 1–20.

Höppe, P. 1999. The physiological equivalent temperature—A universal index for the biometeorological assessment of the thermal environment. *International Journal of Biometeorology* 43 (2): 71–75.

Höppe, P. 2002. Different aspects of assessing indoor and outdoor thermal comfort. *Energy and Buildings* 34 (6): 661–665.

Humphreys, M.A. 1977. The optimum diameter for a globe thermometer for use indoors. *Building Research Establishment Current Paper* 78 (9): 1–5.

Ji, W., B. Cao, M. Luo, and Y. Zhu. 2017. Influence of short-term thermal experience on thermal comfort evaluations: A climate chamber experiment. *Building and Environment* 114: 246–256.

Katavoutas, G., H.A. Flocas, and A. Matzarakis. 2015. Dynamic modeling of human thermal comfort after the transition from an indoor to an outdoor hot environment. *International Journal of Biometeorology* 59 (2): 205–216.

Krüger, E.L., F.O. Minella, and F. Rasia. 2011. Impact of urban geometry on outdoor thermal comfort and air quality from field measurements in Curitiba, Brazil. *Building and Environment* 46: 621–634.

Maruani, T., and I. Amit-Cohen. 2007. Open space planning models: A review of approaches and methods. *Landscape and Urban Planning* 81: 1–13.

Matzarakis, A., and F. Rutz. 2010. *Application of the RayMan Model in Urban Environments.* Freiburg: Meteorological Institute, University of Freiburg.

Nagano, K., A. Takaki, M. Hirakawa, and Y. Tochihara. 2005. Effects of ambient temperature steps on thermal comfort requirements. *International Journal of Biometeorology* 50: 33–39.

Nakayoshi, M., M. Kanda, R. Shi, and R. de Dear. 2015. Outdoor thermal physiology along human pathways: A study using a wearable measurement system. *International Journal of Biometeorology* 59: 503–515.

Nikolopoulou, M., and K. Steemers. 2003. Thermal comfort and psychological adaptation as a guide for designing urban spaces. *Energy and Buildings* 35 (1): 95–101.

Nikolopoulou, M., N. Baker, and K. Steemers. 1999. Improvements to the globe thermometer for outdoor use. *Architectural Science Review* 42: 27–34.

Nikolopoulou, M., N. Baker, and K. Steemers. 2001. Thermal comfort in outdoor urban spaces: Understanding the human parameter. *Solar Energy* 70 (3): 227–235.

Pantavou, K., G. Theoharatos, M. Santamouris, and D. Asimakopoulos. 2013. Outdoor thermal sensation of pedestrians in a Mediterranean climate and a comparison with UTCI. *Building and Environment* 66: 82–95.

Parkinson, T., and R. de Dear. 2015. Thermal pleasure in built environments: Physiology of alliesthesia. *Building Research and Information* 43 (3): 288–301.

Potvin, A. 2000. Assessing the microclimate of urban transitional spaces. In *Proceedings of PLEA2000 (Passive Low Energy Architecture)*, Cambridge, UK, July 2000.

Spagnolo, J., and R. de Dear. 2003. A field study of thermal comfort in outdoor and semi-outdoor environments in subtropical Sydney Australia. *Building and Environment* 38: 721–738.

Taleghani, M., L. Kleerekoper, M. Tenpierik, and A. Dobbelsteen. 2015. Outdoor thermal comfort within five different urban forms in the Netherlands. *Building and Environment* 83: 65–78.

Thorsson, S., F. Lindberg, I. Eliasson, and B. Holmer. 2007. Different methods for estimating the mean radiant temperature in an outdoor urban setting. *International Journal of Climatology* 27: 1983–1993.

Vasilikou, C., and M. Nikolopoulou. 2013. Thermal walks: Identifying pedestrian thermal comfort variations in the urban continuum of historic city centres. In *PLEA2013—29th Conference, Sustainable Architecture for a Renewable Future*, Munich, Germany 10–12 Sept 2013.

Xiong, J., Z. Lian, X. Zhou, J. You, and Y. Lin. 2015. Effects of temperature steps on human health and thermal comfort. *Building and Environment* 94: 144–154.

Yu, Z.J., B. Yang, and N. Zhu. 2015. Effect of thermal transient on human thermal comfort in temporarily occupied space in winter—A case study in Tianjin. *Building and Environment* 93: 27–33.

Yuan, C., and E. Ng. 2012. Building porosity for better urban ventilation in high-density cities—A computational parametric study. *Building and Environment* 50: 176–189.

Chapter 4
Environmental Perception and Outdoor Thermal Comfort in High-Density Cities

Abstract Although outdoor thermal comfort has gained increasing research attention, meteorological conditions and thermal sensation in different urban settings in high-density cities have not been systematically studied from the perspective of urban planning and design. Considering the potential relationship between environmental quality and thermal sensation in outdoor spaces—an emerging topic in perceived comfort, this study offers a new approach for planning and design for climate resilience in cities. This chapter presents the results of an outdoor thermal comfort survey conducted on hot summer days in Hong Kong. Diverse patterns of PET-comfort ratings relationships were found in different urban settings. The study revealed that air temperature, subjective assessments of solar radiation, and wind environment were strong determinants of thermal sensation and evaluation. In our analysis, wind condition showed a significant indirect effect on comfort through subjective perception. Statistical modelling showed that subjective perceptions on microclimate condition and comfort are moderated by various aspects of environmental quality. The findings help inform future design for climate resilience in outdoor urban spaces in hot–humid subtropical cities.

Keywords Outdoor thermal comfort · Thermal sensation · Urban microclimate · Environmental quality · High-density

4.1 Introduction

Outdoor thermal comfort is an integral part of climate-sensitive urban planning and design which improve outdoor thermal comfort by providing shading through urban structures (Thorsson et al. 2011; Lau et al. 2015, 2016), regulating microclimate through vegetation (Bowler et al. 2010; Shashua-Bar et al. 2010), reducing the thermal load (Emmanuel et al. 2007; Lin et al. 2011) and enhancing ventilation which promotes convective heat exchange (Pearlmutter et al. 2007; Krüger et al. 2011; Saneinejad et al. 2014). It also has implications on the physical and mental health of citizens by encouraging the use of outdoor spaces for physical activity and social interactions (Sanches and Pellegrino 2016; Kabisch et al. 2017). Due to the unprecedented rate of urbanisation, citizens increasingly concern the living quality in cities and hence improving the quality of outdoor spaces is one of the important topics in urban planning and design.

4.1.1 Psychological Perception of Environmental Quality

The rapid urban development in sub-tropical and tropical high-density cities has resulted in intense urban heat island effects which considerably affect outdoor thermal comfort (Arnfield 2003). They deteriorate the quality of outdoor spaces which are often perceived as extended living spaces (Ahmed, 2003). Although meteorological factors are decisive to subjective thermal comfort, it is widely recognised that outdoor spaces providing a wider range of thermal conditions and psychological adaptation is important in the outdoor environment because greater fluctuations in thermal conditions facilitate the psychological adaptation of pedestrians (Nikolopoulou and Steemers 2003; Spagnolo and de Dear 2003; Han et al. 2007; Hwang and Lin 2007).

In indoor settings, sensory stimuli have been proved to be associated with thermal comfort. Although hue-temperature hypothesis appears to be questionable (Fanger et al. 1977), studies showed that warm environments are perceived as more comfortable in 5000 K colour temperature light than 2700 K (Candas and Dufour 2005), suggesting that a "cooler" light makes people feel cooler. On the other hand, bright light increases body temperature through the release of melatonin (Badia et al. 1991; Myers and Badia 1993), leading to warmer rating of the environment with bright light. Studies also suggested potential interactions between perceived comfort and perceived air quality and acoustic environments. Clausen et al. (2004) found that a one-degree change in temperature resulted in the same effect on perceived comfort as a 2.4-decipol change in perceived air quality or a 3.9-decibel change in background noise level.

Psychological comfort is generally associated with the ability to see nature (Aries et al. 2010), leading to positive emotions and improved physical comfort (Kaplan 1995). Park et al. (2011) found that the contrasting responses to forests and urban environments significantly influence thermal comfort and that the relationship is bidirectional such that the perception of aesthetics can also be influenced by thermal conditions (Knez 2003; Eliasson et al. 2007). Cultural differences could also modify this relationship since the aesthetic evaluation is highly dependent on cultural and ideological constructs (Knez and Thorsson 2006).

Previous studies suggested that outdoor thermal comfort is associated with perceived environmental quality such as aesthetics (de Castro Fontes et al. 2008), acoustics (Venot and Sémidor 2006), air quality (Pantavou et al. 2017), and perceived control of the environment (Nikolopoulou and Lykoudis 2006). Conventional studies focused on the relationship between subjective thermal sensation and meteorological conditions but largely neglected the environmental stimuli present in outdoor spaces. It leads to a lack of understanding of how the subjective assessment of thermal comfort is influenced by people's perceived environmental quality. It is therefore important for scientists to quantify these interrelationships for enhancing outdoor thermal comfort and hence the use of outdoor spaces in high-density cities.

This study aims to investigate the effect of perceived environmental qualities on subjective thermal perception and how these qualities are associated with outdoor thermal comfort in sub-tropical high-density cities. The potential of mitigating thermal discomfort may be restricted due to the constraints in the physical environment of high-density cities. This study also provides new insights to the design of outdoor spaces by incorporating different environmental qualities for improving outdoor thermal comfort.

4.2 Methodology

4.2.1 High-Density Context of Hong Kong

Hong Kong is located in Southeast Asia at a latitude of 22° 15′ N. It has a sub-tropical monsoon climate and is generally hot and humid in summer. The summer mean air temperature is approximately 28 °C and, in particular, the air temperature in the afternoon often exceeds 31 °C. Relative humidity ranges from 60 and 70% during the daytime in summer.

Hong Kong had a population of 7.4 million in 2017, and the population density in urban areas was approximately 6700 persons per km^2 (Census and Statistics Department 2018). In order to accommodate such a large population, the city is characterised by a high-density and high-rise urban morphology in built-up areas with an average building height of 60 m. The compact urban settings result in intense urban heat island effects and insufficient air ventilation in urban areas.

4.2.2 Micrometeorological Measurements

A field study was conducted in three types of urban settings, including residential areas, urban parks, and streets (Fig. 4.1). There were large variations in environmental conditions due to the urban form, land use, pedestrian activities, and traffic conditions. A total of 13 sites (Fig. 4.2) were selected to conduct micrometeorological measurements with a mobile meteorological station, containing a TESTO 480 data logger for measuring air temperature (T_a), relative humidity (RH), and wind speed (v), and a globe thermometer for measuring globe temperature (T_g) was used in micrometeorological measurements in the resent study. The globe thermometer is composed of a thermocouple wire (TESTO flexible Teflon type K) placed inside a black-painted plastic ball with a diameter (D) of 38 mm and emissivity (ε) of 0.95. Mean radiant temperature (T_{mrt}) was then calculated according to the following equation from Thorsson et al. (2007):

Fig. 4.1 Typical urban outdoor spaces in Hong Kong

$$T_{\mathrm{mrt}} = \left[\left(T_{\mathrm{g}} + 273.15 \right)^4 + \frac{1.10 \times 10^8 * v^{0.6}}{\varepsilon * D^{0.4}} \left(T_{\mathrm{g}} - T_{\mathrm{a}} \right) \right]^{1/4} - 273.15 \qquad (4.1)$$

With the meteorological parameters collected and the respondent's metabolic rate and clothing level, Universal Thermal Climate Index (UTCI) was calculated to represent the objective thermal comfort index. UTCI is commonly used in thermal comfort studies conducted in different climate regions (Krüger et al. 2011; Chan et al. 2017) and sufficiently represents the thermal comfort conditions in objective means.

Fig. 4.2 Location (top) and photos (bottom) of the survey sites

4.2.3 Questionnaire Surveys

Based on previous outdoor thermal comfort studies (Johansson et al. 2014), a questionnaire survey was developed to obtain respondents' subjective sensation of meteorological variables and perception of environmental qualities. It was carried out between June and September 2018, and it took place between 10 and 16 h on weekdays to minimise variations in weather conditions and pedestrian activities. The surveys were conducted at the same location as the mobile meteorological stations. The questionnaire was composed of questions on sensation regarding temperature (TSV), humidity (HSV), wind speed (WSV) and solar radiation (SSV) based on seven-point ASHRAE scale (ASHRAE 2010) such that thermal sensations were reported from cold (−3) to hot (3), with neutral sensation as 0. Overall state of perceived thermal comfort (PCV) was rated on a four-point Likert scale from very uncomfortable (−2) to very comfortable (2) without any option for the neutral state. Perceived environmental qualities, including aesthetic, acoustic, air quality, safety, and convenience, were rated using a five-point Likert scale from very satisfactory (+2) to very unsatisfactory (−2) with a neutral option (0). Demographic background of the participants was also obtained while the activity of the participants before the questionnaire survey (sitting, standing, walking, or doing exercise) was also recorded to represent the metabolic rate. The clothing level was observed by the interviewer using the checklists from ANSI/ASHRAE Standard 55 (ASHRAE 2010).

4.2.4   Statistical Analysis

Logistic regression analysis has been widely used in studies of thermal comfort for understanding how environmental conditions influence thermal perception (Kim et al. 2013; Kumar et al. 2018). The logistic regression equation between the probability of perceiving thermal comfort (p) and perceived environmental quality (T) is defined as:

$$\mathbf{logitp} = \log\left[\frac{\mathbf{p}}{1-\mathbf{p}}\right] = \mathbf{bT} + \mathbf{c} \qquad (4.2)$$

The probability of respondents perceiving thermal comfort can therefore be determined by the odds ratios (ORs) of the significant variables. Statistical analyses using ANOVA, bivariate correlation, and mediation and moderation analysis, were performed to explore the interrelationships between outdoor microclimatic conditions, perceived environmental quality, and sensation of thermal comfort.

4.3   Results and Discussion

4.3.1   Respondent Characteristics

Table 4.1 presents the characteristics of respondents. Male and female respondents accounted for 45.0% and 55.0%, respectively, while young (<18) and old (>55) respondents accounted for about half of the respondents in the present study. Approximately one-third of the respondents were under air-conditioned conditions 15 min before the survey as air-conditioning is relatively common in Hong Kong. The majority of the respondents (53.5%) were walking in the last 15 min and respondents who were standing and sitting 15 min before the survey accounted for 33.2%

Table 4.1 Characteristics of the respondents in the present study

Sex	n	$\%$	15-min AC environment	n	$\%$
Male	863	45.0	Yes	682	35.6
Female	1054	55.0	No	1235	64.4
Age	**n**	**%**	**15-min activity**	**n**	**%**
<18	480	25.0	Sitting	237	12.4
18–24	363	18.9	Standing	637	33.2
25–34	212	11.1%	Walking	1026	53.5%
35–44	175	9.1%	Doing exercise	17	0.9%
45–54	186	9.7%			
>55	491	25.6%	**Total no. of respondents**	1917	
Prefer not revealed	10	0.5%			

and 12.4%, respectively. Only 0.9% of the respondents exercising before conducting the survey since the survey was conducted mostly in street environments during weekdays.

4.3.2   Micrometeorological Measurements

Figure 4.3 shows the boxplots of the meteorological variables measured in the different urban settings. Mean T_a was generally higher in street sites (34.0 °C) due to the high level of pedestrian activities and traffic conditions. Maximum and minimum T_a were also the highest in street sites (38.9 °C and 30.4 °C, respectively). In urban parks or the waterfront, T_a was generally lower, and the temperature range was also smaller, predominantly due to the cooling effect of vegetation and water bodies. As such, RH measured in these sites was also higher in urban parks and the waterfront. Wind speed was relatively consistent across three types of urban settings in the

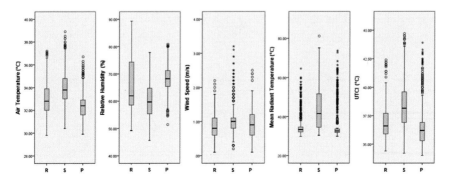

Fig. 4.3 Boxplots of the meteorological variables measured during the survey campaign

present study, with mean v values of approximately 1 m/s. Maximum v measured was slightly higher than 2 m/s with a few outliers reaching up to 3.21 m/s observed in street environment. It is primarily due to the effect of heavy traffic flow during the survey.

The mean T_{mrt} was lower in residential areas and urban parks or the waterfront (35.4 °C and 36.5 °C, respectively), which was due to the presence of extensive vegetation in the areas. The outlier values were observed in the waterfront and open spaces in the housing estates. In the street environment, the complex street geometry resulted in a wide range of radiative environments with a relatively higher average of T_{mrt} (43.9 °C). Due to the predominant influence of T_{mrt}, a similar pattern was observed in UTCI values with the highest mean UTCI values observed in the street environment (38.3 °C). The results show that the urban settings have an influence on the micrometeorological conditions and hence potentially affect subjective thermal perception.

4.3.3 Subjective Thermal Sensation and Perceived Comfort

Spearman's rank correlation coefficients were calculated to evaluate the correlations between measured meteorological parameters and subjective sensations of meteorological conditions for both shaded and unshaded conditions (Table 4.2). For unshaded conditions, TSV was significantly correlated with PET ($r_s = 0.101, p < 0.05$) and T_a ($r_s = 0.150, p = < 0.01$). Results show that subjective sensation of solar radiation (SSV) was the strongest determinant of thermal sensation in unshaded conditions ($r_s = 0.547, p = < 0.001$). Wind sensation (WSV) also shows a significant, negative correlation with thermal sensation ($r_s = -0.185, p < 0.001$). A similar pattern was also observed for perceived comfort (PCV) that highest correlation was found with T_a ($r_s = -0.217, p < 0.001$). PCV was also negatively correlated with SSV ($r_s = -0.343, p < 0.001$) but positively correlated with WSV ($r_s = 0.364, p < 0.001$).

When the respondents were under shaded conditions, the influence of meteorological parameters was more complex as all four measured meteorological parameters were found to be correlated with TSV. It indicates that subjective thermal sensation was no longer dominated by a single meteorological variable. Thermal sensation was also found to be associated with the subjective sensation of solar radiation and wind.

Table 4.2 Bivariate Spearman's rank correlation coefficient between thermal sensation, perceived comfort, and various meteorological parameters (*p < 0.05; **p < 0.01)

		T_a	v	T_{mrt}	PET	SSV	WSV
Unshaded	TSV	**0.150	0.051	0.082	*0.101	**0.547	**−0.185
	PCV	**−0.217	−0.040	−0.021	−0.070	**−0.343	**0.364
Shaded	TSV	**0.159	**0.119	**0.158	**0.142	**0.523	**−0.107
	PCV	**−0.150	−0.004	**−0.152	**−0.156	**−0.237	**0.238

Although the respondents were shaded during the questionnaire survey, SSV still showed the highest correlation with TSV under such typical summer conditions (r_s = 0.523, $p < 0.001$).

4.3.4 Perceived Environmental Quality

Individual acclimatisation, immediate thermal history, and perceived environmental quality can potentially influence subjective thermal sensation and perceived comfort. Moderated regression analysis and corresponding sub-group analysis were conducted to determine the moderators of the relationship between meteorological conditions and subjective thermal sensation, including perceived accessibility, aesthetic quality, acoustic environment, air quality, and perceived safety (Nikolopoulou and Lykoudis 2006). Under unshaded conditions, the effect of air temperature on outdoor thermal comfort was positively moderated by perceived air quality ($F_{(1857)}$ = 3.99, p = 0.021). Sub-group analyses found a stronger relationship between air temperature and outdoor thermal comfort among respondents perceiving good air quality ($p < 0.001$) than those perceiving bad quality (p = 0.284). It was also shown that the sub-group with higher rating on air quality tended to have higher rating on outdoor thermal comfort with the same air temperature (Figs. 4.4 and 4.5).

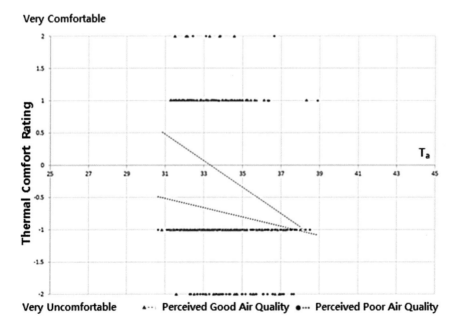

Fig. 4.4 Moderating effect of perceived air quality on the relationship between air temperature and thermal comfort rating

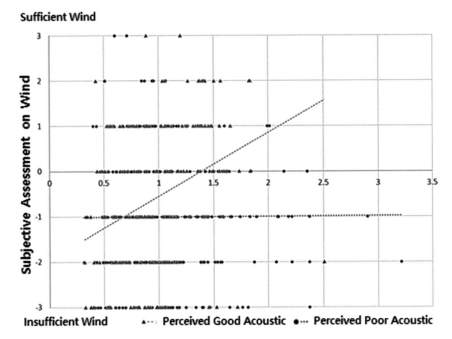

Fig. 4.5 Moderating effect of perceived acoustic environment on the association between wind speed and subjective assessment on wind

Perceived acoustic environment ($F_{(1857)} = 9.39, p = 0.002$) and perceived air quality ($F_{(1857)} = 10.67, p = 0.001$) were significant moderators of the relationship between wind speed and subjective assessment of air movement. The model indicated that when the respondents perceived good acoustic environments, the corresponding subjective assessment of air movement was significantly correlated with wind speed ($p < 0.001$) and had a higher increasing rate with the increase of wind speed. No significant correlation was observed in the relationship between subjective assessment of air movement and wind speed ($p = 0.512$) when the respondents perceived bad acoustic environments. It indicates that increasing wind speed did not result in positive changes in subjective assessment for unsatisfactory acoustic environments.

4.3.5   Overall Thermal Comfort

Logistic regression was used to investigate overall levels of thermal comfort using three sets of variables obtained from the questionnaire surveys and micrometeorological measurements, including measured meteorological variables, sensation of meteorological conditions, and perceived environmental quality. Table 4.3 presents the odds ratios (OR). Crude ORs and ORs adjusted for the three sets of variables were

Table 4.3 Crude and adjusted odds ratios (OR) and 95% confidence intervals (CI) for three sets of variables with respect to respondents reported comfortable votes

	Crude OR	Adjusted OR
Measured meteorological variables		
Wind speed (*v*)	1.102 [0.889, 1.368]	1.178 [0.875, 1.586]
Relative humidity (*RH*)	**1.013 [1.005, 1.021]	*1.017 [1.000, 1.034]
Air temperature (T_a)	1.002 [0.987, 1.017]	1.122 [0.929, 1.355]
Mean radiant temperature (T_{mrt})	1.001 [0.993, 1.009]	1.019 [0.992, 1.047]
UTCI	1.003 [0.989, 1.016]	0.876 [0.713, 1.075]
Subjective sensation		
Thermal sensation (TSV)	***0.246 [0.199, 0.305]	***0.291 [0.226, 0.375]
Solar radiation sensation (SSV)	***2.637 [2.152, 3.231]	***0.537 [0.418, 0.690]
Wind speed (WSV)	***2.938 [2.034, 4.242]	***3.414 [2.262, 5.154]
Humidity sensation (HSV)	***0.490 [0.376, 0.637]	***0.547 [0.409, 0.730]
Perceived environmental qualities		
Convenience	***1.730 [1.336, 2.239]	**1.462 [1.093, 1.956]
Aesthetic	***2.283 [1.867, 2.791]	***1.565 [1.209, 2.025]
Acoustic	***2.063 [1.686, 2.525]	*1.383 [1.073, 1.784]
Air quality	***2.439 [1.981, 3.001]	**1.521 [1.159, 1.996]
Safety	***1.773 [1.380, 2.277]	*1.337 [1.001, 1.785]

calculated for the entire samples ($N = 1971$). Respondents reporting uncomfortable votes were denoted as the reference (OR = 1) in the analyses such that the ORs reported in Table 4.3 represent the probability that the respondents express thermal comfort relative to their counterparts. For example, OR = 2.938 for WSV can be interpreted as respondents who reported comfortable votes being 2.938 times more likely to feel that the air movement is strong (+2) or very strong (+3) than respondents who reported uncomfortable votes.

Respondents expressing thermal comfort were not significantly associated with measured meteorological variables except relative humidity, indicating that thermal comfort in outdoor environments is not necessarily determined by objective means. On the other hand, the crude ORs indicate that subjective sensation of meteorological variables and perceived environmental qualities were significantly associated with human thermal comfort in the outdoor environment. Wind sensation, solar radiation sensation, and perception of air quality have the strongest relationship with thermal comfort. In the adjusted model which took into account all three sets of variables, a similar pattern was observed but most of the effect size was reduced after adjusting for the other variables. This suggests that the interactions between different perceptions of environmental qualities in perceiving comfort in outdoor environment are highly complex and may offset each other in perceiving overall comfort in outdoor spaces.

Among the sensation of meteorological variables, respondents feeling hot (TSV = +2 or +3) were 70.9% less likely to report comfortable votes than the counterparts while respondents feeling strong solar radiation and humidity were, respectively, 46.3% and 45.3% less likely to feel comfortable. However, respondents feeling strong winds were 3.414 times more likely to feel comfortable. It suggests that air movement in dense urban areas is important to outdoor thermal comfort under hot summer conditions. Moreover, satisfaction with aesthetic quality of the outdoor spaces shows the largest effect size among the perceived environmental quality (OR $= 1.565, p = 0.001$). It is closely followed by satisfaction with air quality and convenience with ORs $= 1.521$ and 1.462, respectively ($p = 0.01$).

4.4 Implications on Urban Geometry Design

Previous studies suggested that the most relevant factors to outdoor thermal comfort are T_a and T_{mrt}, which are closely associated with urban geometry, since it is particularly important to provide shading opportunities in urban outdoor spaces. Results of the present study show that although the correlation between T_a and TSV was significant, subjective assessment of meteorological parameters, such as solar radiation and wind speed, showed the highest correlations with subjective thermal sensation. There were also no significant correlations between subjective thermal comfort and measured wind speed due to the weak and relative calm conditions in terms of wind speed in the dense urban environment of Hong Kong (Ng 2009). As such, local respondents were more sensitive to subtle changes in wind speed (Xie et al. 2018).

This study provided further evidence that wind speed plays a moderating role in outdoor thermal comfort by influencing the subjective assessment of air movement. This is particularly important to the urban design of dense urban areas since relatively small improvement of air movement can be perceived by pedestrians to enhance thermal comfort in the outdoor environment. This highlights the need for promoting good practices that improve pedestrian wind environments in high-density cities (Alcoforado et al. 2009; Ng 2009; Reiter 2010).

Our results show that perceived environmental qualities are significantly associated with outdoor thermal comfort. The relationship between outdoor thermal comfort and the degree of positive feeling towards different aspects of perceived environmental qualities was evaluated. Such a positive perception strengthens the contribution of meteorological parameters to subjective assessment of outdoor thermal comfort. Psychological effects associated with negative perception of environmental qualities would suppress the role of co-existing microclimatic factors (Li et al. 2018; McHale et al. 2018). The probability of respondents' reporting comfort votes was also significantly associated with aesthetic, acoustic, air quality, convenience, and safety of the outdoor spaces. This shows the importance of interactions between physical and psychological effects of outdoor thermal comfort (Nikolopoulou and Steemers 2003).

In high-density cities, the limited land resources cause a lot of constraints in the design of the urban geometry, which determines the thermal environment. Design strategies can therefore consider environmental qualities that influence outdoor thermal comfort and increase the adaptive capacity. Further studies are required to investigate how these environmental qualities influence the physiological and psychological pathways of outdoor thermal comfort.

4.5 Conclusions

In this study, the effect of perceived environmental qualities on the relationship between microclimatic variables and subjective assessment of thermal comfort was investigated, based on a questionnaire survey with 1971 responses collected at 13 urban sites. Measured meteorological parameters were not significant in predicting the probability of having comfort votes reported by the respondents. Psychological effect played an important role in the subjective assessment of outdoor thermal comfort since subjective perception of meteorological parameters was found to be significant. The importance of the psychological approach was reiterated by the moderating effects of perceived acoustic and air quality on the relationship between measured and subjective assessment meteorological conditions. In the adjusted logistic regression model, perceived environmental qualities were found to be significant in predicting comfort votes reported by the respondents.

This study provides insights for urban designers to take into account environmental quality in enhancing thermal comfort in the outdoor environment. In high-density cities, the compact urban settings and limited land resources may restrict the potential of enhancing thermal comfort through microclimate so environmental qualities can be effective solutions to enhancing pedestrians' perception of the environment they are located in. Further studies will be required to obtain a more comprehensive understanding of incorporating psychological effect and hence formulate detailed design approaches for climate resilience.

References

Alcoforado, M.J., H. Andrade, A. Lopes, and J. Vasconcelos. 2009. Application of climatic guidelines to urban planning: The example of Lisbon (Portugal). *Landscape and Urban Planning* 90 (1): 56–65.

Aries, M.B.C., J.A. Veitch, and G.R. Newsham. 2010. Windows, view, and office characteristics predict physical and psychological discomfort. *Journal of Environmental Psychology* 30 (4): 533–541.

Arnfield, A.J. 2003. Two decades of urban climate research: A review of turbulence, exchanges of energy and water, and the urban heat island. *International Journal of Climatology* 23: 1–26.

ASHRAE. 2010. *ANSI/ASHRAE Standard 55-2010. Thermal Environmental Conditions for Human Occupancy.* Atlanta: American Society of Heating, Refrigerating and Air-Conditioning Engineers, Inc.

Badia, P., B. Myers, M. Boecker, J. Culpepper, and J.R. Harsh. 1991. Bright light effects on body temperature, alertness, EEG and behaviour. *Physiology & Behaviour* 50 (3): 583–588.

Bowler, D.E., L. Buyung-Ali, T.M. Knight, and A.S. Pullin. 2010. Urban greening to cool towns and cities: A systematic review of the empirical evidence. *Landscape and Urban Planning* 97: 147–155.

Candas, V., and A. Dufour. 2005. Thermal comfort: Multisensory interactions? *Journal of Physiological Anthropology and Applied Human Science* 24 (1): 33–36.

de Castro Fontes, M.S.G., F. Aljawabra, and M. Nikolopoulou. 2008. Open urban spaces quality: A study in a historical square in Bath, UK. In *Proceedings of the 25th Conference on Passive and Low Energy Architecture (PLEA2008)*. October 22–24, 2008, Dublin, Ireland.

Census and Statistics Department. 2018. *Annual Digest of Statistics 2017*. Hong Kong: Census and Statistics Department.

Chan, S.Y., C.K. Chau, and T.M. Leung. 2017. On the study of thermal comfort and perceptions of environmental features in urban parks: A structural equation modeling approach. *Building and Environment* 122: 171–183.

Clausen, G., L. Carrick, P.O. Fanger, S.W. Kim, T. Poulsen, and J.H. Rindel. 2004. A comparative study of discomfort caused by indoor air pollution, thermal load and noise. *Indoor Air* 3 (4): 255–262.

Eliasson, I., I. Knez, U. Westerberg, S. Thorsson, and F. Lindberg. 2007. Climate and behaviour in a Nordic city. *Landscape and Urban Planning* 82 (1): 72–84.

Emmanuel, R., H. Rosenlund, and E. Johansson. 2007. Urban shading—A design option for the tropics? A study in Colombo, Sri Lanka. *International Journal of Climatology* 27 (14): 1995–2004.

Fanger, P.O., N.O. Breum, and E. Jerking. 1977. Can colour and noise influence man's thermal comfort? *Ergonomics* 20 (1): 11–18.

Han, J., G. Zhang, Q. Zhang, J. Zhang, J. Liu, L. Tian, and C. Zheng. 2007. Field study on occupants' thermal comfort and residential thermal environment in a hot-humid climate of China. *Building and Environment* 42 (12): 4043–4050.

Hwang, R.L., and T.P. Lin. 2007. Thermal comfort requirements for occupants of semi-outdoor and outdoor environments in hot-humid regions. *Architectural Science Review* 50 (4): 357–364.

Johansson, E., S. Thorsson, R. Emmanuel, and E. Krüger. 2014. Instruments and methods in outdoor thermal comfort studies—The need for standardization. *Urban Climate* 10: 346–366.

Kabisch, N., M. van den Bosch, and R. Lafortezza. 2017. The health benefits of nature-based solutions to urbanization challenges for children and the elderly—A systematic review. *Environmental Research* 159: 362–373.

Kaplan, S. 1995. The restorative benefits of nature: Toward an integrative framework. *Journal of Environmental Psychology* 15 (3): 169–182.

Knez, I., and S. Thorsson. 2006. Influences of culture and environmental attitude on thermal, emotional and perceptual evaluations of a public square. *International Journal of Biometeorology* 50 (5): 258–268.

Knez, I. 2003. Climate: A nested physical structure in places. In *Proceedings of the 5th International Conference on Urban Climate*, Lodz, Poland.

Krüger, E.L., F.O. Minella, and F. Rasia. 2011. Impact of urban geometry on outdoor thermal comfort and air quality from field measurements in Curitiba, Brazil. *Building and Environment* 46 (3): 621–634.

Lau, K.K.L., F. Lindberg, D. Rayner, and S. Thorsson. 2015. The effect of urban geometry on mean radiant temperature under future climate change: A study of three European cities. *International Journal of Biometeorology* 59 (7): 799–814.

Lau, K.K.L., C. Ren, J. Ho, and E. Ng. 2016. Numerical modelling of mean radiant temperature in high-density sub-tropical urban environment. *Energy and Buildings* 114: 80–86.

Li, Y., D. Guan, S. Tao, X. Wang, and K. He. 2018. A review of air pollution impact on subjective well-being: Survey versus visual psychophysics. *Journal of Cleaner Production* 184: 959–968.

Lin, T.P., R. de Dear, and R.L. Hwang. 2011. Effect of thermal adaptation on seasonal outdoor thermal comfort. *International Journal of Climatology* 31: 302–312.

McHale, C.M., G. Osborne, R. Morello-Frosch, A.G. Salmon, M.S. Sandy, G. Solomon, L. Zhang, M.T. Smith, and L. Zeise. 2018. Assessing health risks from multiple environmental stressors: Moving from G×E to I×E. *Mutation Research/Reviews in Mutation Research* 775: 11–20.

Myers, B.L., and P. Badia. 1993. Immediate effects of different light intensities on body temperature and alertness. *Physiology & Behaviour* 54: 199–202.

Ng, E. 2009. Policies and technical guidelines for urban planning of high-density cities—Air ventilation Assessment (AVA) of Hong Kong. *Building and Environment* 44 (7): 1478–1488.

Nikolopoulou, M., and S. Lykoudis. 2006. Thermal comfort in outdoor urban spaces: Analysis across different European countries. *Building and Environment* 41 (11): 1455–1470.

Nikolopoulou, M., and K. Steemers. 2003. Thermal comfort and psychological adaptation as a guide for designing urban spaces. *Energy and Buildings* 35 (1): 95–101.

Pantavou, K., S. Lykoudis, and B. Psiloglou. 2017. Air quality perception of pedestrians in an urban outdoor Mediterranean environment: A field survey approach. *Science of the Total Environment* 574: 663–670.

Park, B.J., K. Furuya, T. Kasetani, N. Takayama, T. Kagawa, and Y. Miyazaki. 2011. Relationship between psychological responses and physical environments in forest settings. *Landscape and Urban Planning* 102 (1): 24–32.

Pearlmutter, D., P. Berliner, and E. Shaviv. 2007. Integrated modeling of pedestrian energy exchange and thermal comfort in urban street canyons. *Building and Environment* 42 (6): 2396–2409.

Reiter, S. 2010. Assessing wind comfort in urban planning. *Environment and Planning B: Planning and Design* 37 (5): 857–873.

Sanches, P.M., and P.R.M. Pellegrino. 2016. Greening potential of derelict and vacant lands in urban areas. *Urban Forestry and Urban Greening* 19: 128–139.

Saneinejad, S., P. Moonen, and J. Carmeliet. 2014. Coupled CFD, radiation and porous media model for evaluating the micro-climate in an urban environment. *Journal of Wind Engineering and Industrial Aerodynamics* 128: 1–11.

Shashua-Bar, L., O. Potchter, A. Bitan, and Y. Yaakov. 2010. Microclimate modelling of street tree species effects within the varied urban morphology in the Mediterranean city of Tel Aviv, Israel. *International Journal of Climatology* 30: 44–57.

Spagnolo, J., and R. de Dear. 2003. A field study of thermal comfort in outdoor and semi-outdoor environments in subtropical Sydney Australia. *Building and Environment* 38 (5): 721–738.

Thorsson, S., F. Lindberg, I. Eliasson, and B. Holmer. 2007. Different methods for estimating the mean radiant temperature in an outdoor urban setting. *International Journal of Climatology* 27 (14): 1983–1993.

Thorsson, S., F. Lindberg, J. Björklund, B. Holmer, and D. Rayner. 2011. Potential changes in outdoor thermal comfort conditions in Gothenburg, Sweden due to climate change: The influence of urban geometry. *International Journal of Climatology* 31 (2): 324–335.

Venot F., and C. Sémidor. 2006. The soundwalk as an operational component for urban design. In *Proceedings of the 23rd Conference on Passive and Low Energy Architecture (PLEA2006)*. September 6–8, 2006, Geneva, Switzerland.

Xie, Y., T. Huang, J. Li, J. Liu, J. Niu, C.M. Mak, and Z. Lin. 2018. Evaluation of a multi-nodal thermal regulation model for assessment of outdoor thermal comfort: Sensitivity to wind speed and solar radiation. *Building and Environment* 132: 45–56.

Part II
Evaluation of Design Strategies for Outdoor Thermal Comfort

Chapter 5
Effects of Urban Geometry on Mean Radiant Temperature

Abstract Outdoor thermal comfort has been a widely concerned issue in tropical and sub-tropical cities. In order to assess the conditions of outdoor thermal comfort, quantitative information on different spatial and temporal scales is required. The study in this chapter employs a numerical model to examine the spatial and temporal variations of mean radiant temperature (T_{mrt}), as an indicator of radiant heat load and outdoor heat stress in high-density sub-tropical urban environment in summer. The SOLWEIG model is found to simulate the six-directional shortwave and longwave radiation fluxes as well as T_{mrt} very well. Simulation results show that urban geometry plays an important role in intra-urban differences in summer daytime T_{mrt}. Open areas are generally warmer than surrounding narrow street canyons. Street canyons are sheltered from incoming direct solar radiation by shading of buildings, while open areas are exposed to intense solar radiation, especially along the sunlit walls where high T_{mrt} is observed due to reflected shortwave radiation and longwave radiation emitted from the sunlit building walls. The present study confirms that there are great potential in using urban geometry to mitigate high radiant heat load and daytime heat stress in the compacted urban environment. In high-density sub-tropical cities where high daytime T_{mrt} causes severe thermal discomfort in summer, dense urban structures are able to mitigate the extremely high T_{mrt} and improve outdoor thermal comfort. However, the shading strategy has to be cautious about air ventilation in such a dense urban environment.

Keywords High-density · Mean radiant temperature · Sub-tropical · SOLWEIG · Radiant heat load

5.1 Introduction

Urban geometry plays an important role in improving thermal comfort in outdoor environment (Johansson and Emmanuel 2006; Pearlmutter et al. 2007; Andreou 2013). In particular, shading by buildings becomes particularly important in mitigating heat stress in urban areas (Emmanuel and Johansson 2006; Krüger et al. 2011). Previous studies on the effect of urban geometry on outdoor thermal environment are conducted in low- to mid-density urban environment (Ali-Toudert and Mayer 2007; Emmanuel et al. 2007; Pearlmutter et al. 2007; Thorsson et al. 2011; Yahia and Johansson 2013). The understanding of how urban geometry affects outdoor thermal environment in high-density cities is rather limited. It leads to ineffective urban design which may exacerbate thermal discomfort in the dense urban areas.

Mean radiant temperature (T_{mrt}), one of the most important meteorological parameters, governs human energy balance and outdoor thermal comfort (Mayer and Höppe 1987; Ali-Toudert and Mayer 2006). It is defined as the "uniform temperature of an imaginary enclosure in which the radiant heat transfer from the human body equals the radiant heat transfer in the actual non-uniform enclosure" (ASHRAE 2010). In outdoor environment, T_{mrt} is more suitable than conventional parameters like air temperature (T_a) since it shows large spatial variations within short distances, which is particularly important in the complex urban environment. Mayer and Höppe (1987) showed that the difference in T_{mrt} between sun-exposed and shaded locations can reach up to 25 °C at noon, while the difference in T_a is only less than 2 °C, suggesting that T_{mrt} is able to capture the intra-urban difference in outdoor thermal conditions.

This study aims to investigate how urban geometry affects the spatial variation of T_{mrt}, as an indicator of radiant heat load and outdoor heat stress in high-density sub-tropical cities using Hong Kong as a case study. The study was conducted in a central urban area characterised by high-rise buildings and narrow street canyons. The Solar and LongWave Environmental Irradiance Geometry (SOLWEIG) model was employed to simulate T_{mrt} within the study area. Different urban settings, in terms of sky view factor (SVF) and street orientations, were compared, and its implications on the design of urban geometry were also discussed.

5.2   Methodology

5.2.1   Study Area

The study area is located in the central part of Hong Kong and divided into western and eastern halves by a main road (Fig. 5.1). The western side mainly consists of high-rise and densely built commercial buildings and retail activities while the eastern side is mainly composed of less dense but high-rise hotels and commercial buildings with a large open square close to the harbourfront which is located in the south-eastern part of the study area. There is little or no vegetation within the street canyons, despite of the vegetation present along the main road, at the open square and in the park located in the south-western part of the study area.

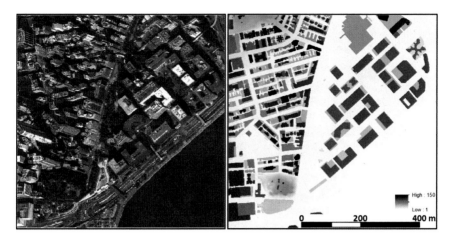

Fig. 5.1 Satellite image (left) and digital surface model (right) of the study area

5.2.2 SOLWEIG Model

SOLWEIG is a MATLAB-based numerical model which simulates spatial variations of 3D radiation fluxes and T_{mrt} in complex urban settings (Lindberg et al. 2008). T_{mrt} is modelled by taking into account the six-directional (upward, downward, and from the four cardinal points) shortwave and longwave radiation fluxes and generating the shadow patterns and sky view factors (SVFs). The model has been proved to provide accurate estimation of the radiation fluxes in different urban settings and weather conditions as well as in different regional contexts (Lindberg and Grimmond 2011). The low computational requirement of the SOLWEIG model allows simulation of larger area and longer period of interest (Lindberg et al. 2008; Lindberg and Grimmond 2011). Therefore, both long-term climate and specific conditions can be simulated. Packages with detailed inputs normally require intensive computational power which limits the use in various situations. As this study aims to investigate the spatial variation of radiant heat load, the SOLWEIG model provides a computationally viable means to obtain information about the spatial variation of T_{mrt} to assess the heat stress in the urban environment.

5.2.3 Input Data and Model Settings

The SOLWEIG model requires two types of input data, namely terrain and meteorological data. Terrain data are in the form of a digital surface model (DSM), including both ground topography and building structures in the study area. The DSM has a spatial resolution of 1 m and is derived from the digital elevation model of Hong Kong as well as building and podium data obtained from the Lands Department.

Podiums are common building structures in Hong Kong that occupy the lower levels of high-rise buildings and generally cover larger building footprint than the building towers, serving as commercial or institutional purposes. Therefore, building tower and podium data are separately documented in Hong Kong. Hourly meteorological data including air temperature, relative humidity, global solar radiation were obtained from two ground-level meteorological stations operated by the Hong Kong Observatory. The diffuse and direct components of solar radiation were calculated according to Lam and Li (1996).

Standard values of absorption coefficients for shortwave and longwave radiation are 0.7 and 0.97 (Lindberg et al. 2016). According to Crawford and Duchon (1999), surface albedo of building and ground surface is designated to 0.2 since the current version model assumes a uniform albedo value over the entire study area. Wind fields and materials of ground and building surfaces are not considered in the current version of the model. The emissivity of building walls was set to be 0.9 since the majority of building surfaces are composed of concrete and glasses while the emissivity of ground surfaces was set to be 0.95 (rough concrete surfaces). Table 5.1 summarises the input values of the parameters used in the present study. Detailed descriptions of the SOLWEIG model are documented in Lindberg and Grimmond (2011) and Lindberg et al. (2008).

5.2.4 Field Measurements

Field measurements were conducted on a clear autumn day (21 October 2014) for assessing the performance of the SOLWEIG model (Fig. 5.2). The instrumental set-up consisted of a TESTO 480 data logger for measuring air temperature, relative humidity and wind speed, and a set of CNR4 net radiometers. Three CNR4 net radiometers were used to measure the shortwave and longwave radiation fluxes from six directions (upward, downward, and the four cardinal points). Tmrt can be determined if the mean radiant flux density (S_{str}) of the human body is known. The six-directional radiation fluxes were multiplied by the angular factors F_i ($i = 1$–6) between a person and the surrounding surfaces using the following equation (Thorsson et al. 2007):

$$S_{str} = \alpha_k \sum_{i=1}^{6} K_i F_i + \varepsilon_p \sum_{i=1}^{6} L_i F_i$$

where K_i and L_i represent the shortwave and longwave radiation fluxes ($i = 1$–6). F_i are the angular factors between a person and the surrounding surfaces ($i = 1$–6). α_k is the absorption coefficient for shortwave radiation with a standard value of 0.7, and ε_p is the emissivity of the human body. According to Kirchhoff's laws, ε_p is equal to the absorption coefficient for longwave radiation with a standard value of 0.97.

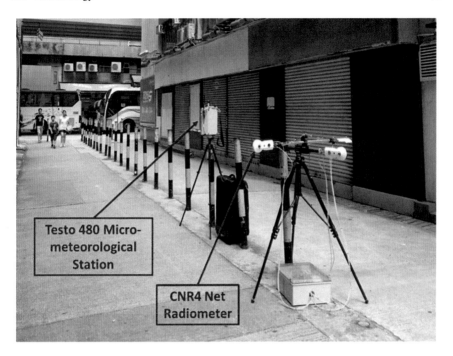

Fig. 5.2 Instrumental settings of the field measurements

For the investigation of heat stress in the study area, a 1D version of SOLWEIG (SOLWEIG1D) was employed to calculate the radiation fluxes and T_{mrt} at a fictitious point (Lindberg 2012). This single point has a user-specified sky view factor (SVF = 0.6) to represent a typical urban setting and is assumed to be sunlit during daytime (Lindberg et al. 2014). However, this is impossible in reality when SVF is smaller than 1. In the present study, the spatial maps of the hours when T_{mrt} is higher than 60 °C at this point were selected and subsequently averaged to obtain the spatial variation of T_{mrt} under such heat-stress conditions.

5.3 Validation of the SOLWEIG Model

Shortwave and longwave radiation fluxes obtained from field measurements were compared to the modelled results. Figure 5.3 shows that the SOLWEIG model simulated both shortwave and longwave radiation fluxes well (R^2-value > 0.9). The modelling of shortwave radiation fluxes was more straightforward because the three components of shortwave radiation were used as the input for the SOLWEIG model. However, the reflected shortwave component was a simplified estimation so the

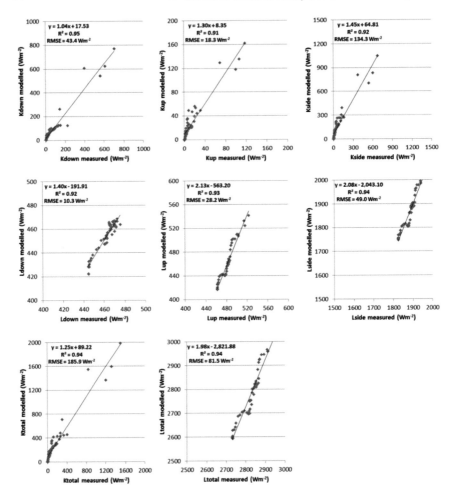

Fig. 5.3 Correlation between modelled and measured all the shortwave and longwave radiation fluxes at the measurement site

estimation of shortwave radiation fluxes from the lateral directions was less accurate. Given the location of the validation site in an N-S street canyon, the estimation of eastward and westward shortwave radiation showed lower R^2-values (0.80 and 0.83, respectively) due to the reflection from building surfaces. The root mean square errors (RMSE) for the downward ($K_\downarrow$), upward ($K_\uparrow$), side (K_{side}), and total (K_{total}) shortwave fluxes were 43.4, 18.3, 134.3, and 185.9 Wm^{-2} respectively. The underestimation of downward and southward shortwave radiation is believed to be associated with the different sky conditions of the measurement sites and meteorological stations. Nonetheless, it captured the diurnal variation of both shortwave and longwave radiation fluxes considerably well.

Estimating longwave radiation fluxes was more complex because their estimation is largely based on other meteorological parameters and empirical constants (Lindberg et al. 2008). The SOLWEIG model generally simulated the six-directional longwave radiation fluxes well, with R^2-values over 0.9 for downward and upward fluxes and 0.85 for lateral directions. RMSEs for the downward ($K_\downarrow$), upward ($K_\uparrow$), side (K_{side}), and total (K_{total}) shortwave fluxes were 10.3, 28.2, 49.0, and 81.5 Wm^{-2} respectively. When the street canyon was sunlit, the model tended to overestimate longwave fluxes by about 20 Wm^{-2} (40 Wm^{-2} for southward direction). On the other hand, it generally underestimated longwave fluxes by up to 40 Wm^{-2} when the street canyon was shaded by surrounding buildings.

The scatterplot between modelled and measured values of T_{mrt} at 15-min interval showed high correlation with an R^2-value of 0.93 (Fig. 5.4). Modelled values were overestimated by about 10 °C at noon due to the different sky conditions between the measurement site and the meteorological station at around noon. It was overestimated by 5–7 °C in the early afternoon because the southward and westward longwave radiation was overestimated, which was also found by Lindberg et al. (2008) at denser urban location where longwave radiation was emitted from surrounding building walls, affecting a significant part of human energy balance. It also agreed with Ali-Toudert's (2005) findings that more than 70% of the energy is absorbed by a standing body in the form of longwave radiation during daytime and demonstrates the importance of accurately simulating longwave radiation and estimating T_{mrt} by field measurements.

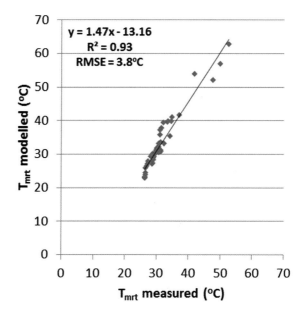

Fig. 5.4 Correlation between modelled and measured T_{mrt} at 15-min interval at the study site

$y = 1.47x - 13.16$
$R^2 = 0.93$
RMSE = 3.8 °C

5.4 Spatial Variation of Mean Radiant Temperature

5.4.1 Outdoor Mean Radiant Temperature in Different Urban Settings

The SOLWEIG model was capable of simulating the spatial variation of three-dimensional radiation fluxes and hence the T_{mrt} of a large spatial extent. The summer average of the spatial variation of daytime T_{mrt} of the study area is shown in Fig. 5.5. T_{mrt} was generally higher (>35 °C) in open areas during daytime, including the large open square, parks and areas along the main road. It was even higher (near 40 °C) in harbourfront areas due to the intense direct solar radiation reaching the building and ground surfaces, as well as subsequent reflection of shortwave and longwave radiation. It also showed in the higher T_{mrt} in front of the sunlit walls at the square (3–4 °C higher than areas in front of the north-east-facing walls). The higher level of direct solar radiation absorbed by the ground also led to a substantial level of T_{mrt}, as a result of increased longwave radiation exposed to a standing body (Masmoudi and Mazouz 2004). Lower T_{mrt} was more confined to the north-east- and north-west-facing walls where shading is more common. Open areas had average daytime T_{mrt}

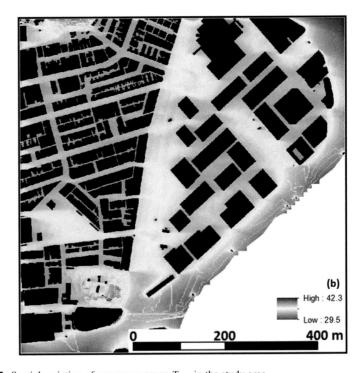

Fig. 5.5 Spatial variation of summer average T_{mrt} in the study area

up to 37 °C during summer, indicating that intense radiant heat load and heat stress
are likely to occur.

Within street canyons, daytime T_{mrt} was considerably lower (up to 5 °C lower than
the open area) due to the denser urban structures preventing directing solar radiation
reaching the building and ground surfaces, resulting in a deeply shaded street canyon.
It was also shown by the difference in T_{mrt} between wide and narrow street canyons
located in the eastern and western halves of the study area, respectively. T_{mrt} was
about 2 °C higher in the wider street canyons (SVF $\approx$ 0.6), suggesting that denser
urban structures are able to mitigate heat stress by effective shading of surrounding
buildings. There were considerable differences between north–south (N-S) and east–
west (E-W) street canyons. T_{mrt} is higher in E-W canyons especially on the sunlit
side of the canyon. On the other hand, N-S canyons exhibited lower T_{mrt} due to the
shading of the buildings on the western side. This was particularly prominent in the
street intersections, where T_{mrt} was higher than the shaded parts of N-S canyons,
suggesting that N-S canyons can mitigate high T_{mrt} and improve heat stress at the
pedestrian level.

5.4.2 Heat-Stress Area

The one-dimensional version of the SOLWEIG model was used to determine hours
when a fictitious point with user-specified SVF (0.6) experiences T_{mrt} over 60 °C so
as to examine areas experiencing severe heat stress. Figure 5.6 shows the spatial variation of severe heat stress by averaging the spatial maps of the days with these hours,
which are normally common in summer and early autumn. High T_{mrt} was found in
areas close to sunlit walls facing south-east to south-west, especially within E-W

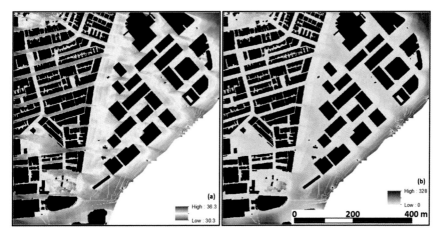

Fig. 5.6 a Averaged hourly T_{mrt} and **b** frequency of T_{mrt} over 60 °C when T_{mrt} of the fictitious
point is higher than 60 °C

street canyons (Fig. 5.6a). This is due to the intense direct and reflected shortwave radiation as well as longwave radiation emitted from surrounding sunlit surfaces. T_{mrt} was about 1.5 °C lower on the southern side of the E-W canyons but confined to areas in front of the building walls where it was shaded most of the time. Heat stress was not observed in N-S canyons due to shading of buildings throughout the day.

In open areas, hot spots were found at the southern corners of buildings where a large amount of reflected shortwave and longwave radiation emitted from the sunlit building surfaces was experienced. Heat stress was also observed along south-west-facing building walls in the open square. However, hotspots were spatially limited to areas close to building walls. It is somewhat different from those observed in street canyons due to the more open setting which allowed the more rapid release of longwave radiation to surrounding areas. Longwave radiation emitted from the sunlit walls was reflected by the opposite side of the canyon, resulting in stronger heat stress within street canyons. The frequency of hours when heat stress ($T_{mrt} > 60$ °C) occurred is shown in Fig. 5.6b. It was found that the occurrence of heat stress was more frequent in highly open areas like harbourfront. This resulted in unfavourable conditions in these leisure areas where urban dwellers engage in outdoor activities.

5.5 Temporal Variation of Mean Radiant Temperature

Figure 5.7 shows the temporal variation of T_{mrt} at four selected locations (E-W canyon, N-S canyon, square, and courtyard). High T_{mrt} (above 40 °C) was generally observed from July to September due to the high solar altitude during daytime. Within the E-W canyon, such high T_{mrt} was observed during most of the daytime from 10 to 16 h due to the prolonged exposure to solar radiation. On the other hand, high T_{mrt} was more confined from 11 to 14 h within the N-S canyon since the high-rise buildings provided extensive shading in the morning and afternoon. In addition, the difference between shaded and sunlit hours can be up to 19.2 °C. In winter months, T_{mrt} was maintained at a relatively higher level (30–36 °C) in the N-S canyon than the E-W canyon. It provided a certain level of solar access in the N-S canyon which were favourable to pedestrians.

In more open areas, high T_{mrt} up to 50 °C were found from 11 to 17 h in the summer. In the square, much longer periods of high T_{mrt} were found throughout the day and also lasted from March to November, indicating that intense heat stress is more common and intense in open squares. However, minimum T_{mrt} was lower than that in the street canyons due to the release of longwave radiation during night-time. Similar maximum T_{mrt} was observed in the courtyard although there was a shorter period of such high T_{mrt} occurred (3–4 h at noon). Compared to the square, high T_{mrt} was only observed from May to September and the temporal pattern resembled that of the N-S canyon.

Figure 5.8 illustrates the changes in the spatial variation of average hourly T_{mrt} in times of heat stress ($T_{mrt} > 60$ °C) from 11 to 15 h. At 11 h, hot spots were generally found in open areas such as the harbourfront areas. They were unlikely to be found on

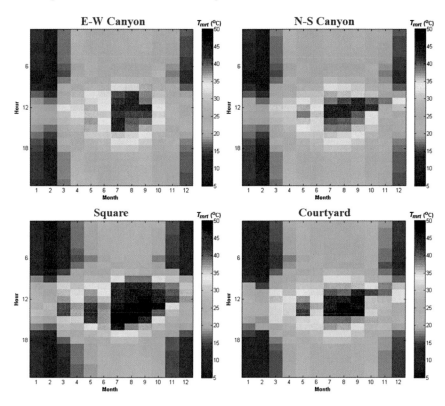

Fig. 5.7 Temporal variation of mean radiant temperature at four urban locations

the northern side of buildings due to their shading. As the sun progressively rises to the highest elevation of the day, cool areas were limited in front of the north-east- or north-west-facing walls of the buildings. T_{mrt} was generally higher than that observed in the morning. The hottest time of the day was found to be at 14 h, and hot spots were found in front of sunlit building walls because of longwave radiation emitted from the sunlit walls which absorbed a large amount of shortwave radiation under clear-sky conditions. The blockage of the cooler part of the sky by buildings also explains such a pattern (Lindberg 2012). The effect of shading is best demonstrated in the spatial variation of T_{mrt} at 15 h. Cooler areas were mainly found in shaded areas on the eastern side of the buildings. The extent of shaded areas covered almost the entire main road in the middle of the study area except the junctions with E-W streets on its western side. This suggests that shading is considerably effective in mitigating daytime heat stress in the complex urban environment. It can be enhanced by a denser urban form with increasing building height where appropriate (Thorsson et al. 2011; Andreou 2013). The use of shading of the buildings on the southern side of the E-W streets should also be maximised in order to mitigate heat stress (Andreou 2014). For example, increasing the width of pedestrian pathways on the

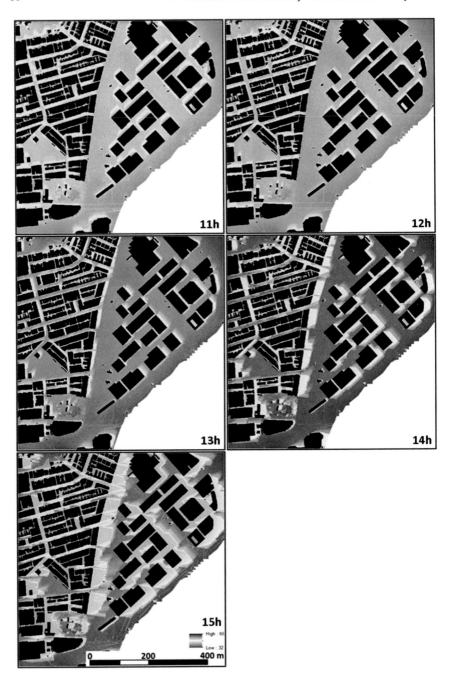

Fig. 5.8 Average hourly T_{mrt} when T_{mrt} of the fictitious point is higher than 60 °C from 11 to 15 h

southern side allows pedestrians to stay under shade when the sun is at the highest altitude.

In order to develop guidelines for urban design with respect to outdoor thermal comfort, other factors such as ventilation and vegetation should be taken into account as well. As a coastal city, sea breezes are often considered as one of the important measures to improve urban microclimates in Hong Kong (Ng 2009). Ventilation in dense urban areas is largely dependent on site coverage ratio, building separation, and height (Yuan and Ng 2012). Therefore, it is important to provide sufficient ventilation and, at the same time, avoid excessive solar heating in the urban areas. Further investigation is required to determine an "optimal" urban geometry according to local microclimatic characteristics. Shading can also be provided by vegetation with dense tree crowns (Shashua-Bar et al. 2006; Krayenhoff et al. 2014). However, the availability of land for vegetation is generally limited in the dense urban areas of high-density cities. Vegetation should be provided for locations where heat stress is observed. The findings of the present study provide information about where vegetation is potentially required.

5.6   Conclusions

This study used the SOLWEIG model to examine the spatial variation of T_{mrt} in high-density sub-tropical urban environment. The model performed well and successfully simulated the six-directional shortwave and longwave radiation fluxes as well as T_{mrt}. Though further field measurements are required to better validate the model under different meteorological conditions, T_{mrt} was found to be higher in open areas than street canyons due to the exposure to direct shortwave radiation. High T_{mrt} was also observed in E-W street canyons, especially on the south-facing walls as a result of reflected shortwave radiation and longwave radiation emitted from the building walls. N-S canyons are therefore preferred in order to mitigate high T_{mrt}. Areas along sunlit building walls were also found to be hot spots where the highest T_{mrt} was observed. Shading was found to be an effective measure to mitigate high T_{mrt} as the lowest T_{mrt} was generally observed in the shaded areas along east-facing walls.

Findings of the present study suggest that shading of buildings can improve radiant heat load at street level. However, the shading strategy has to be cautious about air ventilation in such a dense urban environment. Instead of increasing building height and density, using artificial shading devices such as overhanging façades and extended canopies from buildings is preferred in order to enhance pedestrian-level thermal comfort. Mitigation measures also include vegetation which lowers T_{mrt} through shading and evapotranspiration. Further work of the present study includes field measurements for better validation of the models under different meteorological conditions and the examination of the effect of different sky conditions on the spatial variation of T_{mrt} in the urban environment.

References

Ali-Toudert, F., and H. Mayer. 2006. Numerical study on the effects of aspect ratio and orientation of an urban street canyon on outdoor thermal comfort in hot and dry climate. *Building and Environment* 41: 94–108.

Ali-Toudert, F., and H. Mayer. 2007. Thermal comfort in an east–west oriented street canyon in Freiburg (Germany) under hot summer conditions. *Theoretical and Applied Climatology* 87: 223–237.

Ali-Toudert, F. 2005. *Dependence of Outdoor Thermal Comfort on Street Design*. Freiburg: Berichte des Meteorologischen Institutes der Universität Freiburg. November 2005, No. 15.

Andreou, E. 2013. Thermal comfort in outdoor spaces and urban canyon microclimate. *Renewable Energy* 55: 182–188.

Andreou, E. 2014. The effect of urban layout, street geometry and orientation on shading conditions in urban canyons in the Mediterranean. *Renewable Energy* 63: 587–596.

ASHRAE. 2010. *ANSI/ASHRAE Standard 55-2010. Thermal Environmental Conditions for Human Occupancy*. Atlanta: American Society of Heating, Refrigerating and Air-Conditioning Engineers, Inc.

Crawford, T.M., and C.E. Duchon. 1999. An improved parameterization for estimating effective atmospheric emissivity for use in calculating daytime downwelling longwave radiation. *Journal of Applied Meteorology* 38: 474–480.

Emmanuel, R., and J. Johansson. 2006. Influence of urban morphology and sea breeze on hot humid microclimate: The case of Colombo, Sri Lanka. *Climate Research* 30 (3): 189–200.

Emmanuel, R., H. Rosenlund, and E. Johansson. 2007. Urban shading—A design option for the tropics? A study in Colombo, Sri Lanka. *International Journal of Climatology* 27: 1995–2004.

Johansson, E., and R. Emmanuel. 2006. The influence of urban design on outdoor thermal comfort in the hot, humid city of Colombo, Sri Lanka. *International Journal of Biometeorology* 51: 119–133.

Krayenhoff, E.S., A. Christen, A. Martilli, and T.R. Oke. 2014. A multi-layer radiation model for urban neighbourhoods with trees. *Boundary-Layer Meteorology* 151 (1): 139–178.

Krüger, E.L., F.O. Minella, and F. Rasia. 2011. Impact of urban geometry on outdoor thermal comfort and air quality from field measurements in Curitiba, Brazil. *Building and Environment* 46 (3): 621–634.

Lam, J.C., and D.H.W. Li. 1996. Correlation between global solar radiation and its direct and diffuse components. *Building and Environment* 31 (6): 527–535.

Lindberg, F., and C.S.B. Grimmond. 2011. The influence of vegetation and building morphology on shadow patterns and mean radiant temperatures in urban areas: Model development and evaluation. *Theoretical and Applied Climatology* 105: 311–323.

Lindberg, F., B. Holmer, and S. Thorsson. 2008. SOLWEIG 1.0—Modelling spatial variations of 3D radiant fluxes and mean radiant temperature in complex urban settings. *International Journal of Biometeorology* 52: 697–713.

Lindberg, F., B. Holmer, S. Thorsson, and D. Rayner. 2014. Characteristics of the mean radiant temperature in high latitude cities—Implications for sensitive climate planning applications. *International Journal of Biometeorology* 58 (5): 613–627.

Lindberg, F., S. Thorsson, D. Rayner, and K. Lau. 2016. The impact of urban planning strategies for reducing heat stress in a climate change perspective. *Sustainable Cities and Society* 25: 1–12.

Lindberg, F. 2012. *The SOLWEIG-Model*. Sweden: University of Gothenburg. Available from: http://www.gvc.gu.se/Forskning/klimat/stadsklimat/gucg/software/solweig/

Masmoudi, S., and S. Mazouz. 2004. Relation of geometry, vegetation and thermal comfort around buildings in urban settings, the case of hot arid regions. *Energy and Buildings* 36: 710–719.

Mayer, H., and P. Höppe. 1987. Thermal comfort of man in different urban environments. *Theoretical and Applied Climatology* 38: 43–49.

Ng, E. 2009. Policies and technical guidelines for urban planning of high-density cities—Air ventilation assessment (AVA) of Hong Kong. *Building and Environment* 44: 1478–1488.

Pearlmutter, D., P. Berliner, and E. Shaviv. 2007. Integrated modeling of pedestrian energy exchange and thermal comfort in urban street canyons. *Building and Environment* 42: 2396–2409.

Shashua-Bar, L., M.E. Hoffman, and Y. Tzamir. 2006. Integrated thermal effects of generic built forms and vegetation on the UCL microclimate. *Building and Environment* 41: 343–354.

Thorsson, S., F. Lindberg, I. Eliasson, and B. Holmer. 2007. Different methods for estimating the mean radiant temperature in an outdoor urban setting. *International Journal of Climatology* 27 (14): 1983–1993.

Thorsson, S., F. Lindberg, J. Bjorklund, B. Holmer, and D. Rayner. 2011. Potential changes in outdoor thermal comfort conditions in Gothenburg, Sweden due to climate change: The influence of urban geometry. *International Journal of Climatology* 31: 324–335.

Yahia, M.W., and E. Johansson. 2013. Influence of urban planning regulations on the microclimate in a hot dry climate: The example of Damascus, Syria. *Journal of Housing and the Built Environment* 28: 51–65.

Yuan, C., and E. Ng. 2012. Building porosity for better urban ventilation in high-density cities—A computational parametric study. *Building and Environment* 50: 176–189.

Chapter 6
Urban Greening Strategies for Enhancing Outdoor Thermal Comfort

Abstract Hong Kong suffers from an intense urban heat island effect of up to 4 °C as a result of compact urban form and highly urbanised land cover. Enhancing the cooling efficiency of urban greenery is essential for improving the microclimate in high-density cities. This paper aims to delineate design strategies for urban greenery to maximise thermal benefits and mitigate the daytime UHI effect. Two site-specific design strategies for tree planting in the urban environment are proposed. The sky view factor-based design approach and the wind-path design approach are evaluated in the neighbourhood scale in two climate-sensitive areas with different urban morphologies. Observed data and simulation results indicated that the cooling effect of urban trees is highly associated with SVF. Air temperature reduction (a 1.5 °C reduction) is the most profound for the high-SVF scenario, whereas substantial radiation shading (T_{mrt} reduced to 34 °C) is detected in areas with medium–low SVFs. The modelling study also showed that the cooling of air temperature and sensible heat were twice as high for vegetation arranged in wind corridors than those for leeward areas. The study demonstrated that tree planting in conjunction with proper planning is an effective measure to mitigate daytime UHI.

Keywords Urban greenery · Urban heat island mitigation · High-density

6.1 Introduction

The vulnerability of cities to climate change increases due to high population density and rapid urban growth, especially in high-density cities like Hong Kong where urban heat island (UHI) intensity in Hong Kong ranges between 2 and 4 °C (Siu and Hart 2013). Such high UHI intensity results in severe heat stress which triggers health problems to the urban inhabitants (Goggins et al. 2012). A previous study in Hong Kong reported that, for a 1 °C rise above 29 °C, there is a 4.1% increase in natural mortality in areas with high UHI intensity, compared to only 0.7% in areas with low UHI intensity (Goggins et al. 2012). Thermal discomfort and heat stress also affect the livelihood of vulnerable groups such as elderly people and those with chronic diseases (Luber and McGeehin 2008; Chau et al. 2009).

Urban greenery is widely regarded as an effective measure to mitigate UHI in urban areas and improve the urban microclimate. At the macro scale, intra-urban measurements showed that there was a 4 °C difference in air temperature between the urban core and a highly vegetated area in the city. Armson et al. (2012) also reported that urban greenery effectively cooled urban surfaces by 20 °C and reduced globe temperature by 5–7 °C. At building block scale, a 10% increase in the greenery ratio resulted in 0.8 °C of cooling (Dimoudi and Nikolopoulou 2003). Observational study further showed that urban parks were 1 °C cooler than surrounding non-vegetated sites (Bowler et al. 2010). The microclimatic benefits of urban greenery are determined by foliage density (Theodosiou 2003) and leaf area density (LAD) and Leaf Area Index (LAI) were proved to be important metrics in defining the cooling benefits of different tree species (Spangenberg et al. 2008). A previous field study in hot and humid climate showed that air and surface temperatures were reduced by 1.3 °C and 14.7 °C, respectively, with a maximum LAD of $1.0m^2/m^3$ and LAI value of 5 (Shinzato and Duarte 2012). Numerical modelling also showed that tree canopies effectively intercept 84% of horizontal direct radiation with a maximum LAD of $1.8m^2/m^3$ and LAI value of 3 (Fahmy et al. 2010).

The sky view factor (SVF) has been shown as a key parameter for the impact of building morphology on urban microclimate. A previous study investigated the relationship between intra-urban temperature differences and areal SVF means, and it was found that area-averaged SVF values predicted temperature between sites better than point-based SVF values (Unger 2009). Another study in Hong Kong reported that a 1% reduction in SVF was found to reduce the daytime UHI intensity by 1–4% (Giridharan et al. 2004). Chen et al. (2012) also found that the influence of the SVF on daytime air temperature increase in urban sites varied with building density. Moreover, previous studies also suggested that SVF was negatively associated with the effective emissivity of an urban canopy due to multiple scattering and refection caused by building structures (He et al. 2015).

In hot, humid climates, it is important for urban designers to minimise thermal discomfort and reduce the UHI effect in densely built urban areas. Previous studies have been mostly conducted in dry climates under clear sky conditions (Fahmy et al. 2010). However, there is a lack of systematic research on climate-sensitive urban planning and design for hot, humid climates. In sub-tropical regions, the hot, humid climate during summer results in partially cloudy conditions. It is therefore important to obtain a comprehensive understanding of how urban greenery performs under different sky conditions in order to provide solutions for urban planners and designers to optimise the planning of tree planting.

In high-density cities like Hong Kong, there is a requirement of 30% coverage of urban greenery as a mitigation measure to UHI for new developments (Ng et al. 2012). However, it is challenging to achieve this requirement due to the city's limited land resources and compact urban morphology. As such, optimising the design of urban greenery by taking into account the built environment and microclimatic conditions can enhance the thermal benefits of urban greenery in compact urban environments (McPherson et al. 1994; Shashua-Bar and Hoffman 2004). Hence, further studies on context-based planning and design strategies of urban greenery in high-density urban environment are urgently needed. This study employs numerical modelling to evaluate two site-specific design approaches of urban trees on mitigating daytime UHI in a compact urban environment. The cooling effect of different greenery design strategies at site level was investigated. Urban planners and designers can therefore maximise the benefits of urban greenery to improve daytime thermal comfort and mitigate the UHI effect in high-density urban environments.

6.2 Methodology

6.2.1 Study Areas for Tree Design Approaches

Two morphology-oriented design strategies were considered for tree planting in high-density urban areas of Hong Kong. A morphology-based approach was proposed for urban areas with irregular building layout and different building heights, while a wind-path approach was proposed for areas with regular building block arrays and a prevailing wind direction in summer. A sensitivity test was conducted to evaluate these two design approaches in two urban areas, namely Tsim Sha Tsui (TST) and Sham Shui Po (SSP), respectively, because of their representative urban morphology and local climate characteristics (Fig. 6.1).

The TST area has mixed land uses with an irregular distribution of tall commercial towers and residential buildings. Total building volume in the area is approximately 11,718,000 m^3 with a mean and standard deviation of building height of 37 m and 24 m, respectively. The SVF values at ground level range from 0.2 to 0.8. The

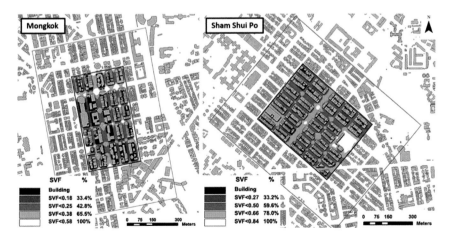

Fig. 6.1 Locations of the selected sites with building layout and spatial distribution of SVF for **a** MK and **b** SSP for the investigation of tree design approaches

morphology-based method was evaluated using the morphology of the TST area. The cooling effects of the following design scenarios were evaluated:

- Trees arranged in spots with low SVF (<0.2);
- Trees arranged in spots with medium SVF (0.2–0.4); and
- Trees arranged in spots with high SVF (0.4–0.8).

The SSP area is a traditional residential district with regular building block array geometry. Total building volume in the area is 16,722,000 m^3, with a mean and standard deviation of building height of 32 m and 15 m, respectively. This area is next to the waterfront, and the main streets in the study area are parallel or 45° to the prevailing summer wind from the waterfront. A wind-path approach for tree arrangement was evaluated in the SSP area. The thermal performance of trees in the following design scenarios was analysed:

- Trees arranged in streets parallel to summer wind direction (along wind path);
- Trees arranged in streets perpendicular to summer wind direction (at sheltered position); and
- Trees arranged in streets aligned 45° to summer wind direction (lower wind speed).

With reference to the Hong Kong Planning Standard and Guidelines, a greenery coverage ratio of 25% was adopted in this study. The influence of trees on air temperature (T_a) and surface temperature (T_s), as well as the radiant and convective environments, were compared to evaluate the performance of mitigating daytime UHI. Mean radiant temperature (T_{mrt}) was widely recognised as a key parameter to the thermal conditions in the outdoor environment in hot climates (Lin et al. 2010; He et al. 2015) so the radiative cooling was also analysed to evaluate different design approaches.

6.2.2 Field Measurements

Field measurements were conducted to evaluate the thermal benefits or trees under low and high SVF (SVF ranging from 0.2 to 0.8). The trees selected in this study had similar solar transmissivity ratios, which are determined by the ratio of downward shortwave radiation under a canopy and at an exposed location (measured by a thermopile-type pyranometer). Microclimatic variables were measured under the tree canopy and at an exposed reference point at a height of 1.5 m. The mobile meteorological station contained a HOBO sensor and Testo400 instruments was placed at each measurement point to record the downward shortwave radiation, air temperature, relative humidity, and wind speed at a 10^{-s} sampling interval (Fig. 6.2). Globe temperature was also measured by a standard globe thermometer with a diameter (D) of 0.15 m and emissivity (ε) of 0.95. T_{mrt} was then calculated using the following equation (Thorsson et al. 2007):

| Testo400 Multi-purpose Instruments | Kipp & Zonen Radiation Indicator | FLIR Thermography Camera | Mobile Meteorological Station |

Fig. 6.2 Instruments and mobile meteorological station used in this study

$$T_{\mathrm{mrt}} = \left[\left(T_g + 273.15\right)^4 + \frac{1.1 \times 10^8 v_a^{0.6}}{\varepsilon D^{0.4}} \times \left(T_g - T_a\right) \right]^{1/4} - 273.15$$

The measurements were conducted from 12 h 3 0min to 14 h under both clear and cloudy conditions in July and August 2014. Several criteria were established to ensure the representativeness of the data acquired, including: (1) the background solar radiation within the same range for each weather condition (clear sky: 800–1000 W m^{-2} of global solar radiation; cloudy sky: 350–400 W m^{-2} of diffuse solar radiation); (2) the solar radiation within a constant range 30–60 min before measurements and during the measurement period; and (3) weak wind conditions (wind speed lower than 1.5 ms^{-1}).

6.2.3 Simulation Settings

The simulation settings adopted a 250 × 250 × 30 grid for the model domain. The grid size for the TST and SSP models was set as 3 m and 6 m, respectively, for the investigation of design approaches. For the study of tree-planting strategies, a grid size of 3 m was used for the models (MK and SSP). Dense vertical telescoping grids (three layers) were set for the first 2 m to achieve more accurate results since this study focuses on the thermal environment at the pedestrian level. Table 6.1 describes the input parameters of the model. Most buildings in the city centre of Hong Kong are decades old with dark-coloured façades so the albedo values of the building surfaces

Table 6.1 Input data for the ENVI-met model

Initial air temperature [K]	303	Relative humidity at 2 m [%]	70
Factor of shortwave adjustment	1.2	Cloud cover	2/8
Heat transmission walls [W/m^2 K]	2	Heat transmission roof [W/m^2 K]	2
Albedo walls	0.2	Albedo roofs	0.3

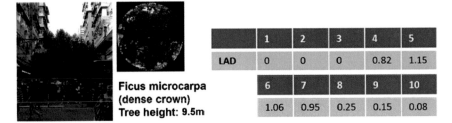

	1	2	3	4	5
LAD	0	0	0	0.82	1.15
	6	7	8	9	10
	1.06	0.95	0.25	0.15	0.08

Ficus microcarpa (dense crown) Tree height: 9.5m

Fig. 6.3 *Ficus microcarpa* is popular species in Hong Kong (photo taken in the SSP area)

in the model were set to low values (0.2 and 0.3 for walls and roofs, respectively). Initial T_a and RH were determined from the mean values of the record from hot sunny days in August 2012 (based on the ground-level meteorological stations at Sham Shui Po and Hong Kong Observatory Headquarter). The solar radiation adjustment factor was set to 1.2 to simulate the intense radiation level on a clear summer day in Hong Kong. Sub-tropical regions usually have cloudy summers (Giridharan et al. 2008), and a small number of high-level clouds are common under sunny weather. As such, a 2/8 fraction of high clouds was set for the simulation. Since both study sites are located in waterfront areas, sea breezes and its directional effects are common phenomena in local climates. Input wind data, including wind direction and pedestrian wind velocity, can be extrapolated down from the 500-m-high wind data of the district by using the wind profile power law expression.

The vegetation of the model was profiled according to a 10-m-high *Ficus microcarpa* planted in the SSP area (Fig. 6.3). *F. microcarpa* is a native species that is widely planted in urban areas of Hong Kong and has been frequently proposed in roadside tree-planting plans (Planning Department 2003). The maximum value in the LAD profile of the tree model is around 1.2 m^2/m^3, and the LAI value of the tree model is approximately 4.5 (Ali-Toudert and Mayer 2006). The shortwave albedo of the leaf is assumed to have the default value of 0.2. It has been shown that urban greenery provides substantial cooling during daytime (Cheng et al. 2012). The study aimed to evaluate the effect of urban trees on mitigating daytime UHI so the simulation period was set from 08 to 16 h.

The simulation period was from 06 to 14 h, and the results between 13 and 14 h were extracted for analysis. Apart from T_a, surface temperature (T_s) of ground surface and T_{mrt} were calculated to evaluate the outdoor radiative environment (Givoni et al. 2003; Armson et al. 2012). PET was calculated using the BioMet package of the ENVI-met model to evaluate the effects of roadside trees on pedestrian comfort. For the settings of human parameters in BioMet, the clothing index was adjusted to 0.4 to represent the average level of clothing in summer Hong Kong (Givoni et al. 2003; Cheng et al. 2012). Metabolic activity was set to 90 Wm^{-2} to represent standing or light activities in the outdoor environment (Havenith et al. 2012).

6.3 Results and Discussion

6.3.1 Field Measurements Under Different Sky Conditions

Results show that the net effects of trees are associated with SVF. When trees with similar solar transmission ratios are individually planted at urban areas with similar SVF values, cooling outcomes in terms of T_{mrt} are comparable under similar levels of background solar radiation. Reduction in T_{mrt} achieved by trees at sites with SVF values of 0.52 and 0.50 was 21.7 °C and 21.6 °C, respectively, given that the solar transmission ratio of tree crown was approximately 0.08. Reduction in trees with solar transmission ratios of 0.06 at sites with SVF values of 0.73 and 0.70 was 17.5 °C and 17.0 °C, respectively.

There were significant differences in the net thermal benefits of trees planted under low- and high-SVF conditions for both clear and cloudy conditions. Under clear sky conditions, the reductions in T_{mrt} under and high and low SVF were approximately 30 °C and 23 °C, respectively. For cloudy conditions, the level of reduction was 15 °C under high SVF and 3.0 °C under low SVF (Fig. 6.4). Larger cooling magnitude in T_{mrt} was observed at high-SVF locations than low-SVF locations, indicating that the cooling effects of urban trees were more significant on sunny days.

On the other hand, the influence of building morphology was more evident on cloudy days. Under cloudy conditions, diffuse shortwave radiation is more uniform over the sky compared to clear sky conditions in the early afternoon (Noorian et al. 2008). Therefore, with the same range of background solar radiation and tree canopy transmissivity, the thermal benefit of trees in terms of T_{mrt} reduction is dependent on the fraction of the sky obstructed by tree crowns under cloudy conditions. Since cloudy weather is common in summer sub-tropical regions due to the increased

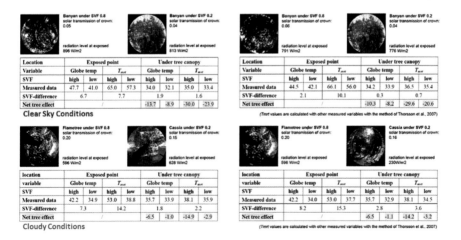

Fig. 6.4 Comparison between reductions in globe temperature and T_{mrt} under different SVFs in sunny and cloudy conditions

occurrence of low-level clouds, morphology-based planning strategies are important for the design strategies of roadside tree planting in sub-tropical cities (Field and Barros 2014).

6.3.2 Validation and Sensitivity of the ENVI-Met Model

PET, T_{mrt}, and T_s obtained from the field measurements were compared to the modelled results (Fig. 6.5). It showed that the ENVI-met model simulated PET and T_{mrt} reasonably well (R^2-values of 0.73 and 0.82, respectively), indicating that it is a reliable tool for studying the thermal benefits of trees on urban microclimates in sub-tropical regions.

The sensitivity of the ENVI-met model to surface albedo showed that the variations in T_{mrt} were attributed to changes in wall albedo for both high- and low-SVF cases. Under cloudy conditions, the differences in T_{mrt} between high and low wall albedo values were approximately 3–4 °C for the high-SVF case, depending on the distance from the wall surface. The magnitude of difference was larger under sunny conditions with a range of 5–8 °C for the low-SVF case due to the change in the

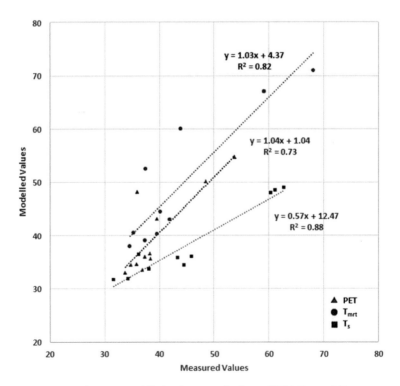

Fig. 6.5 Comparison between modelled and measured values of PET, T_{mrt} and T_s

incoming solar radiation. This indicates that the influence of surface characteristics was more prominent in dense urban environment. Under clear sky conditions, the difference in T_a associated with wall albedo alterations was approximately 0.5 °C for the high-SVF case. For the low-SVF case, it was close to 0.8 °C next to wall surfaces and 0.6–0.7 °C in the street. These results are consistent with the findings of Taha (1997) and Sailor (1995). The sensitivity tests demonstrated the significant effect of surface albedo on the microclimate variables in the model, particularly affecting the modelled T_{mrt} values. As most of the buildings in the study areas are decades old with dark-toned façades, the albedo of building surfaces was adjusted to lower values of 0.2 and 0.3 for walls and roofs, respectively (Ng 2001; Buildings Department 2004).

6.3.3 Morphology-Based Approaches

The morphology-based approach of tree planting was evaluated in the TST area. Spatial correlation was observed between SVF and the spatial pattern of T_s and T_a for the base case (Fig. 6.6). The contours of T_s and T_a were more closely spaced when there was an abrupt change in SVF (as shown by the colour scale), indicating that SVF is a significant indicator of urban microclimatic conditions. In the first scenario, the trees were placed in the open spaces with high-SVF values ranging from 0.4 to 0.8. This type of green matrix is very common in street parks in the dense urban areas. As these high-SVF areas were exposed to intense solar radiation, shading provided by trees created substantial cooling, as shown by the reduction in T_s by 10 °C and 13 °C in the morning and afternoon, respectively. This is more prominent at noon as T_s was decreased from 49 to 31 °C. T_{mrt} under tree canopy was reduced by 27 °C compared to the base-case scenario, and these values were very similar to the observed values at noon in the sensitivity test. In the model, the turbulent fluxes were calculated as the results of the wind shearing and thermal stratification. The

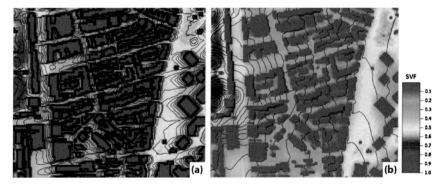

Fig. 6.6 Spatial pattern of SVF comparing to spatial distribution of **a** surface temperature and **b** air temperature

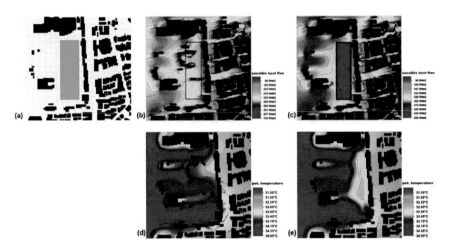

Fig. 6.7 **a** Large patch of trees located in areas with high SVF. Sensible heat flux **b** without and with **c** trees, and air temperature at 1.5 m **d** without and **e** with trees at 13 h

upward sensible heat flux from the ground reaches 380 Wm^{-2} at noon for the base case, while the sensible heat fluxes became negative because of the presence of trees in the vegetated scenario under the same weather condition (Fig. 6.7). It suggests that the tree canopy is able to maintain the ground surface cooler than the air above. Moreover, as the air under the tree canopy was under shade for a long period of time, T_a at pedestrian level was lower by 1 °C and 1.5 °C in the morning and afternoon, respectively.

Similar number of trees were arranged in urban spaces with medium SVF ranging from 0.2 to 0.4 in the second scenario. Such an arrangement is commonly found in areas with medium-sized, connected green spaces. Modelled results show that considerable cooling was observed when trees were located in areas with medium SVF. Ts of the shaded ground surface was cooled by 15 °C and 5 °C at 13 h and 16 h, respectively, which is consistent with Armson et al. (2012). The tree canopy significantly altered the longwave radiation budget of ground surface, and the radiation released from ground surface was reduced from 20–40 Wm^{-2} to 6 Wm^{-2} (Fig. 6.8). T_{mrt} was reduced by 23 °C, and reduction in T_a was less significant in the medium-SVF scenario (approximately 0.3 °C). The two planting scenarios show that the most significant cooling was found at street intersections (Shashua-Bar and Hoffman 2000).

In the third scenario, trees were arranged in areas with low SVF (<0.2). This arrangement was normally found in isolated pocket parks in the inner urban areas, especially in commercial districts and old urban cores. The cooling effect was similar to that of the medium-SVF scenario although modification of T_s and overall radiation budget were slightly lower in the low-SVF scenario because of the shading effect of the surrounding compact urban geometry.

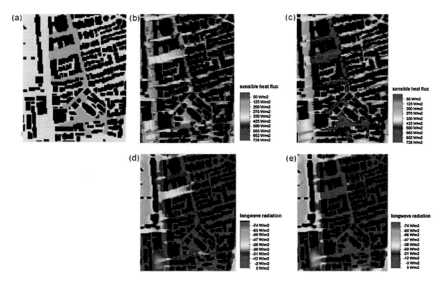

Fig. 6.8 **a** Connected green corridors in medium-SVF areas. Sensible heat flux **b** without and with **c** trees, and longwave radiation budget **d** without and **e** with trees at 13 h

These case studies demonstrate that the microclimatic benefits of trees are associated with urban morphology (i.e. SVF). In urban areas with irregular building layout, the morphology-based approach provides a solution for planners to reduce outdoor thermal stress by using appropriate arrangements of street trees. Centralised arrangements of trees allow for the cooling effect to be accumulated and achieve substantial reduction in T_a, T_s, and T_{mrt} in areas with high SVF during daytime. Previous studies demonstrated that urban morphology has a significant influence on the radiative environment and outdoor thermal conditions in low-latitude cities (Ali-Toudert and Mayer 2007; Emmanuel et al. 2007). As such, the net shading effect varies with the locations of trees under different SVF, which was also shown in Armson et al. (2012) and Andreou (2014). This study found that substantial in T_s and T_{mrt} can be achieved by placing trees in areas with low-to-medium SVF during early afternoon due to the high solar altitude in sub-tropical summer. Cheng et al. (2012) suggested that the comfortable range of T_{mrt} is between 32 and 34 °C in urban areas of sub-tropical cities in summer. The sensitivity test indicated that street trees created comfortable microclimate by reducing T_{mrt} to approximately 34 °C in some of the heavily built areas with low SVF.

6.3.4 Wind-Path Approach

The SSP area was selected to evaluate the wind-path approach of tree planting. As shown in the modelled results (Fig. 6.9), there were considerable differences in

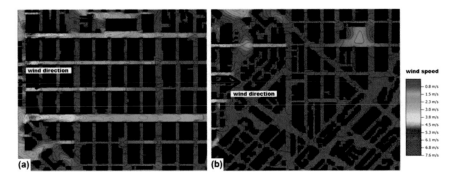

Fig. 6.9 Modelled wind field in the base case of the SSP area. Wind speed was about 2–4 ms^{-1} in streets aligned with wind direction and was reduced to less than 1 ms^{-1} in leeward and diagonal streets

wind speed among the streets at 0°, 45°, and 90° to the prevailing wind direction. Wind speed in diagonal (45°) and leeward streets was 13% and 10% of that aligned with the prevailing wind direction, respectively. In the first scenario, trees were placed in the main streets aligned with the prevailing wind and diagonal streets two blocks away. The net effect of tree planting is shown in Fig. 6.10 by comparing the base case and vegetated scenario. Remarkable cooling was observed when the trees were placed along the wind paths and T_a was reduced by 0.6 to 0.8 °C at

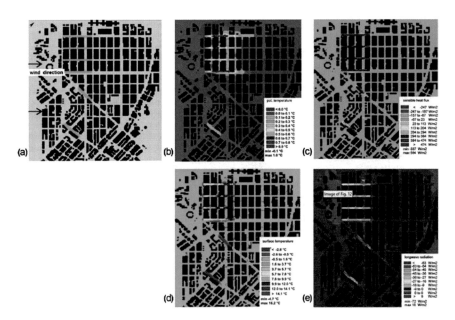

Fig. 6.10 a Street trees arranged in wind paths and diagonal streets. Cooling magnitude on **b** air temperature, **c** sensible heat flux, **d** surface temperature, and **e** longwave radiation at 13 h

street intersections at noon. Cool air was extended to the leeward areas adjacent to the main streets by approximately 30 m into the downwind areas. Similar levels of cooling were also reported by Jauregui (1991) and Crewe (2003). The cooling effect was smaller in the diagonal streets (0.2 °C reduction in T_a). Synergetic effects were also detected in sensible heat reduction when the effect of street trees was combined with air ventilation along wind paths. The reduction in sensible heat due to trees in the streets aligned with prevailing wind direction was twice that resulted from trees in the diagonal streets. The cooling effect in terms of T_s and T_{mrt}, sensitive to the surrounding urban morphology (Ali-Toudert and Mayer 2007), was found to be less dependent on wind direction. The modelled reduction of T_{mrt} reached 27 °C and 30 °C in the wind paths and diagonal streets, respectively.

In the second scenario, the same number of trees was located in the streets perpendicular to the wind and in the leeward diagonal streets a few blocks away. The reduction in T_a and sensible heat flux was considerably smaller when the trees were placed in the leeward areas than that in the first scenario (Fig. 6.11). T_a in the leeward streets was decreased by 0.3 °C at noon and by 0.2 °C in the morning and afternoon. Bruse and Fleer (1998) concluded that the cooling effect of trees was reduced with decreasing wind speed. Hong Kong is a coastal city with substantial onshore wind in summer. In waterfront areas where a regular street layout is common, the channelling effect can be achieved in streets with small angles to the prevailing wind (Amorim

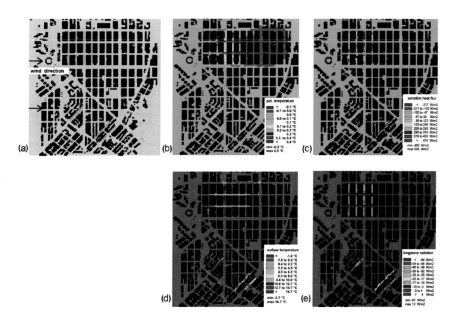

Fig. 6.11 **a** Street trees arranged in leeward and diagonal streets. Cooling magnitude on **b** air temperature, **c** sensible heat flux, **d** surface temperature, and **e** longwave radiation at 13 h

et al. 2012; Claus et al. 2012). Results of this study also show that remarkable synergetic effect can be obtained when trees were located in the wind paths and the cooling effect on T_a and sensible heat was increased by a factor of two and three.

6.4   Conclusions

This study investigated optimised tree-planting strategies for mitigating daytime UHI in high-density sub-tropical cities. Morphology-based and wind-path approaches were evaluated in two climate-sensitive waterfront areas in Hong Kong. It shows that small parks in high-SVF areas with highly localised tree planting significantly cool down T_a and mitigate daytime UHI effect in urban areas. On the other hand, medium-size green spaces in medium-SVF areas and small green fractions under low-SVF areas effectively reduce the radiative load in low-latitude cities in the early afternoon, such that relatively comfortable T_{mrt} are provided in heavily built-up areas. In compact areas with limited land availability for green spaces, tree planting along wind paths is recommended to enhance the cooling benefits in the neighbourhood.

The major limitation of this study was the use of a single-tree model in evaluating the cooling effect so further studies are required to investigate the thermal performance of different tree species so that systematic evaluation and appropriate selection of tree species can be achieved for a better planning of urban greening. Moreover, the analysis of tree-planting approaches was limited to the impacts of urban trees on microclimate regulations. Cobenefits of urban trees also include ecological functions which support ecosystems in the urban environment. Further work shall also focus on optimising planning strategies in relation to multiple functions of urban greening in the built environment.

References

Ali-Toudert, F., and H. Mayer. 2006. Numerical study on the effects of aspect ratio and orientation of an urban street canyon on outdoor thermal comfort in hot and dry climate. *Building and Environment* 41: 94–108.

Ali-Toudert, F., and H. Mayer. 2007. Effects of asymmetry, galleries, overhanging façades and vegetation on thermal comfort in urban street canyons. *Solar Energy* 81: 742–754.

Amorim, J., J. Valente, C. Pimentel, A. Miranda, and C. Borrego. 2012. Detailed modelling of the wind comfort in a city avenue at the pedestrian level. In *Usage, Usability, and Utility of 3D City Models, EDP Sciences*, 03008.

Andreou. 2014. The effect of urban layout, street geometry and orientation on shading conditions in urban canyons in the Mediterranean. *Renewable Energy* 63: 587–596.

Armson, D., P. Stringer, and A. Ennos. 2012. The effect of tree shade and grass on surface and globe temperatures in an urban area. *Urban Forestry & Urban Greening* 11 (3): 245–255.

Bowler, D.E., L. Buyung-Ali, T.M. Knight, and A.S. Pullin. 2010. Urban greening to cool towns and cities: A systematic review of the empirical evidence. *Landscape and Urban Planning* 97 (3): 147–155.

Bruse, M., and H. Fleer. 1998. Simulating surface-plant-air interactions inside urban environments with a three-dimensional numerical model. *Environmental Modelling and Software* 13: 373–384.

Buildings Department. 2004. *A Guideline of Building Regulation Submission for Lighting Requirements of Buildings Using Computational Simulations*, Technical Report for Buildings Department. CAO G55:35p.

Chau, P.H., K.C. Chan, and J. Woo. 2009. Hot weather warning might help to reduce elderly mortality in Hong Kong. *International Journal of Biometeorology* 53 (5): 461–468.

Chen, L., E. Ng, X. An, C. Ren, M. Lee, U. Wang, and Z. He. 2012. Sky view factor analysis of street canyons and its implications for daytime intra-urban air temperature differentials in high-rise, high-density urban areas of Hong Kong: A GIS-based simulation approach. *International Journal of Climatology* 32: 121–136.

Cheng, V., E. Ng, C. Chan, and B. Givoni. 2012. Outdoor thermal comfort study in a subtropical climate: A longitudinal study based in Hong Kong. *International Journal of Biometeorology* 56: 43–56.

Claus, J., O. Coceal, T.G. Thomas, S. Branford, S. Belcher, and I.P. Castro. 2012. Wind-direction effects on urban-type flows. *Boundary-Layer Meteorology* 142 (2): 265–287.

Crewe, K. 2003. *The Potential of Climate Modelling in Greenway—Planning for Phoenix*. Arizona.

Dimoudi, A., and M. Nikolopoulou. 2003. Vegetation in the urban environment: Microclimatic analysis and benefits. *Energy and Buildings* 35 (1): 69–76.

Emmanuel, R., H. Rosenlund, and E. Johansson. 2007. Urban shading—A design option for the tropics? A study in Colombo, Sri Lanka. *International Journal of Climatology* 27 (14): 1995–2004.

Fahmy, M., S. Sharples, and M. Yahiya. 2010. LAI based trees selection for mid latitude urban developments: A microclimatic study in Cairo, Egypt. *Building and Environment* 45 (2): 345–357.

Field, C.B., and V.R. Barros. 2014. *Climate Change 2014 Impacts, Adaptation, and Vulnerability*.

Giridharan, R., S. Ganesan, and S. Lau. 2004. Daytime urban heat island effect in high-rise and high-density residential developments in Hong Kong. *Energy and Buildings* 36 (6): 525–534.

Giridharan, R., S. Lau, S. Ganesan, and B. Givoni. 2008. Lowering the outdoor temperature in high-rise high-density residential developments of coastal Hong Kong: The vegetation influence. *Building and Environment* 43 (10): 1583–1595.

Givoni, B., M. Noguchi, H. Saaroni, O. Pochter, Y. Taacov, N. Feller, and S. Becker. 2003. Outdoor comfort research issues. *Energy and Buildings* 35 (1): 77–86.

Goggins, W.B., E.Y.Y. Chan, E. Ng, C. Ren, and L. Chen. 2012. Effect modification of the association between short-term meteorological factors and mortality by urban heat islands in Hong Kong. *PLOS One* 7(6): e38551.

Havenith, G., D. Fiala, K. Błazejczyk, M. Richards, P. Bröde, I. Holmer, H. Rintamaki, Y. Benshabat, and G. Jendritzky. 2012. The UTCI-clothing model. *International Journal of Biometeorology* 56 (3): 461–470.

He, X., S. Miao, S. Shen, J. Li, B. Zhang, Z. Zhang, and X. Chen. 2015. Influence of sky view factor on outdoor thermal environment and physiological equivalent temperature. *International Journal of Biometeorology* 59 (3): 285–297.

Jauregui, E. 1991. Influence of a large urban park on temperature and convective precipitation in a tropical city. *Energy and Buildings* 15 (3): 457–463.

Lin, T., A. Matzarakis, and R. Hwang. 2010. Shading effect on long-term outdoor thermal comfort. *Building and Environment* 45 (1): 213–221.

Luber, G., and M. McGeehin. 2008. Climate change and extreme heat events. *American Journal of Preventive Medicine* 35 (5): 429–435.

McPherson E.G., D.J. Nowak, and R.A. Rowntree. 1994. *Chicago's Urban Forest Ecosystem: Results of the Chicago Urban Forest Climate Project*. General Technical Report No. NE-186. Radnor, PA: U.S. Department of Agriculture, Forest Service, Northeastern Forest Experiment Station: 201p.

Ng, E. 2001. A study on the accuracy of daylighting simulation of heavily obstructed buildings in Hong Kong. *Building Simulation* 7: 1215–1222.

Ng, E., L. Chen, Y. Wang, and C. Yuan. 2012. A study on the cooling effects of greening in a high-density city: An experience from Hong Kong. *Building and Environment* 47: 256–271.

Noorian, A.M., I. Moradi, and G.A. Kamali. 2008. Evaluation of 12 models to estimate hourly diffuse irradiation on inclined surfaces. *Renewable Energy* 33 (6): 1406–1412.

Planning Department. 2003. *Stage II Study on Review of Metroplan and the Related Kowloon Density Study Review.*

Sailor, D.J. 1995. Simulated urban climate response to modifications in surface albedo and vegetative cover. *Journal of Applied Meteorology* 34 (7): 1694–1704.

Shashua-Bar, L., and M. Hoffman. 2000. Vegetation as a climatic component in the design of an urban street: An empirical model for predicting the cooling effect of urban green areas with trees. *Energy and Buildings* 31 (3): 221–235.

Shashua-Bar, L., and M.E. Hoffman. 2004. Quantitative evaluation of passive cooling of the UCL microclimate in hot regions in summer, case study: Urban streets and courtyards with trees. *Building and Environment* 39 (9): 1087–1099.

Shinzato, P., and D. Duarte. 2012. Microclimatic effect of vegetation for different leaf area index-LAI. In *PLEA2012 Conference*, November 7–9, 2012, Lima, Peru.

Siu, L.W., and M.A. Hart. 2013. Quantifying urban heat island intensity in Hong Kong SAR, China. *Environmental Monitoring and Assessment* 185 (5): 4383–4398.

Spangenberg, J., P. Shinzato, E. Johansson, and D. Duarte. 2008. Simulation of the influence of vegetation on microclimate and thermal comfort in the city of São Paulo. *REVSBAU* 3 (2): 1–19.

Taha, H. 1997. Modelling the impacts of large-scale albedo changes on ozone air quality in the south coast air basin. *Atmospheric Environment* 31 (11): 1667–1676.

Theodosiou, T.G. 2003. Summer period analysis of the performance of a planted roof as a passive cooling technique. *Energy and Buildings* 35 (9): 909–917.

Thorsson, S., F. Lindberg, I. Eliasson, and B. Holmer. 2007. Different methods for estimating the mean radiant temperature in an outdoor urban setting. *International Journal of Climatology* 27 (14): 1983–1993.

Unger, J. 2009. Connection between urban heat island and sky view factor approximated by a software tool on a 3D urban database. *International Journal of Environment and Pollution* 36 (1): 59–80.

Chapter 7
Effect of Tree Species on Outdoor Thermal Comfort

Abstract Tree planting is one of the veritable tools for combating urban heat island and improving thermal comfort in the wake of global warming and urbanisation. However, trees of different species and morphological properties have variable solar attenuation capacity and consequently, thermal comfort regulation potential. Besides, the shadow-cast effect by buildings helps in reducing pedestrian radiant load and consequently improves thermal comfort, especially in high-density cities even though ventilation is reduced. Therefore, a holistic and contextual understanding of tree planting and shadow-casting can help in designing climate-proof cities. In this study, we employed the ENVI-met model to better understand the interaction between these two forms of shading (trees and buildings) on the pedestrians' thermal comfort in Hong Kong and the influence of one over the other. The impact of different urban densities on the thermal comfort improvement potential by eight common tree species in Hong Kong was specifically studied. Results show that shallow canyons are susceptible to worse thermal condition when compared to their deeper counterparts with similar aspect ratio value. Of all tree configuration parameters, leaf area index, tree height, and trunk height are most influential in improving and aggravating daytime and night-time thermal comfort, respectively. We also found that trees' effectiveness in improving daytime thermal comfort reduces with increasing urban density and vice versa for night-time. For the reference of planners and landscape architects, this study recommends tall trees of low canopy density with high trunk in deeper canyons and vice versa for shallow canyons and open areas.

Keywords ENVI-met · Thermal comfort · Tree planting · Tree species · Street canyon · Urban density

7.1 Introduction

Trees are commonly used in urban areas to create a thermally comfortable environment owing to the extensive shading and evaporative cooling capacity of trees, especially in tropical and subtropical climates (Bowler et al. 2010; Gómez-Baggethun and Barton 2013. It also lowers the air and surface temperature under tree canopies

© The Author(s), under exclusive license to Springer Nature Singapore Pte Ltd. 2022   101
K. K.-L. Lau et al., *Outdoor Thermal Comfort in Urban Environment*,
SpringerBriefs in Architectural Design and Technology,
https://doi.org/10.1007/978-981-16-5245-5_7

(McPherson et al. 2011; Armson et al. 2012) and hence reduces heat storage and convection (Matzarakis et al. 2010; Armson et al. 2013). The shading effect of trees leads to the attenuation of mean radiant temperature (T_{mrt}), a key determinant of human thermal comfort, at street level (Tan et al. 2013). The degree of the attenuation of solar radiation and T_{mrt} varies across tree species due to their physical characteristics such as leaf area density, tree height, crown diameter, coverage ratio, and separation distance (Berry et al. 2013). On the other hand, water vapour is discharged from trees through their leaf stomata via transpiration (Oke et al. 1989). It regulates the latent heat loss during the phase change of water from liquid state to water vapour, resulting a cooler leaf surface and surrounding environment. The transpiration rate is dependent on local environmental factors (air temperature and humidity, soil moisture content) and physical configuration of the trees (tree height, leaf thickness and colour, trunk and branch architecture) (Fahmy et al. 2010; de Abreu-Harbich et al. 2015). Selection of tree species is therefore important to the implementation of mitigation strategies in local neighbourhoods.

The setting of tree-planting sites is another important determinant of the microclimatic modification and enhancement of thermal comfort. Previous studies were mostly conducted in more open settings such as urban parks, residential, or institutional environments, with only few focusing on urban street canyons which can be defined by aspect ratio (AR) or sky view factor (SVF). SVF is defined as the fraction of sky visible from a certain point at pedestrian level while AR refers to the ratio between the bounding building height and the street width. The relationship between the aspect ratio of street canyon, microclimatic benefits, UHI intensity has been extensively investigated in previous studies (Ali-Toudert and Mayer 2006; Qaid and Ossen 2014; Theeuwes et al. 2014; Morakinyo and Lam 2016) and can be explained by two counteracting processes (Theeuwes et al. 2014). The first process is the shadowing effect during daytime, which blocks the exposure to direct solar radiation and prevents warming of the street canyon. Shadow canyons (i.e. AR $\leq$ 1) allow solar exposure to most of the wall and ground surfaces, resulting in a large amount of energy transfer between canyon surfaces and overlying air masses. It warms up the air within street canyons and, combining with the intense radiant load, impacts human thermal comfort. In contrast, deeper street canyons (i.e. AR $\geq$ 2), where high-rise buildings impede solar access, increase average coverage of shadow-cast, reduce surface and radiant energy, provide improved thermal comfort conditions during daytime. The second mechanism dominates at night-time as emitted heat (in the form of longwave radiation) from building and ground surfaces is trapped within street canyons. Such trapped energy further increases the air temperature of the street canyons and accentuates UHI, especially in deeper canyons.

In subtropical cities like Hong Kong, tree shading is important to reduce the thermal load by minimising direct exposure to solar radiation although trees may impede airflows. The magnitude of such effects depends on tree species due to their physical configurations and morphological characteristics (Fahmy et al. 2010; de Abreu-Harbich et al. 2015; Morakinyo and Lam 2016). As such, recent studies have shifted towards the thermal effect of specific tree species and its implications on mitigation strategies and decision-making. This study therefore aims to investigate

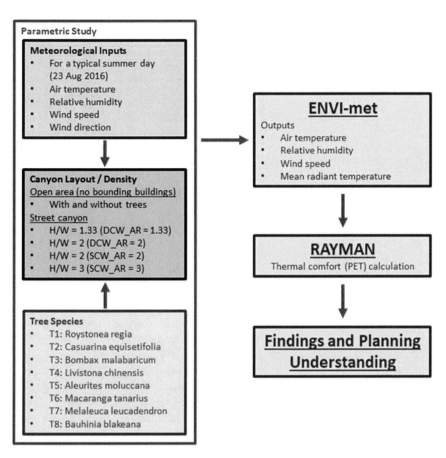

Fig. 7.1 Methodological framework of this study

the effect of common tree species in Hong Kong by using numerical modelling. It determines the potential for improvement of thermal comfort by selected tree species and the magnitude of thermal benefits during daytime and night-time by tree shading and shadow-casting in different urban contexts. It specifically addresses the shadow-casting effect of in-canyon trees. The implications on tree-planting strategies in high-density urban planning and greenery projects are also discussed.

7.2 Methodology

7.2.1 Study Framework

The methodological framework of this study is presented in Fig. 7.1. The influence of shadow-cast and tree shade of eight selected species planted in four street canyons with different aspect ratios and a reference case was investigated in the parametric study. Findings provide a comprehensive understanding of the interrelationships between tree planting, shadow-casting, and the microclimatic benefits in order to develop strategies of tree selection in a real neighbourhood in Hong Kong. Numerical modelling was carried out using ENVI-met V4.0 and RayMan1.2 models for a typical summer day.

7.2.2 Model Description and Initialisation

Two microclimate models, namely ENVI-met V4.0 and RayMan1.2, were used in the present study. ENVI-met is a holistic three-dimensional non-hydrostatic model based on the principles of computational fluid dynamics for modelling plant-surface-atmosphere interactions in complex environment with buildings of different shapes, height and materials, road or surface of different materials, and vegetation of different configuration. It can simulate with very high spatial (up to 0.5 m horizontally) and temporal resolution (up to 10 s) which enables near-accurate modelling of microclimatic parameters (Bruse and Fleer 1998; Huttner et al. 2009). Vegetation is not only presented as a porous media to solar radiation and airflow, but also treated as a biological body which interacts with the surrounding environment by evapotranspiration (which is a function of leaf temperature and aerodynamic resistance, and leaf-to-air humidity deficit). Various coefficients of obstructions, according to lead density distribution and the height of plants, were estimated in the ENVI-met model to take into account attenuation of direct and diffuse shortwave radiation and atmospheric and terrestrial longwave radiation (Bruse and Fleer 1998; Morakinyo and Lam 2016). Detailed descriptions of the ENVI-met model were documented in previous studies (Bruse and Fleer 1998; Huttner et al. 2009) while Salata et al. (2016) provided a comprehensive sensitivity study. RayMan 1.2 was developed for estimating spatial and time-dependent radiation flux densities, mean radiant temperature, and thermal comfort indices around complex urban structures, with urban morphology parameters and vegetation data available. In addition, it can estimate the above thermal parameter where only point meteorological data are available.

Hourly averaged meteorological variables such as air temperature (T_a), relative humidity (RH), wind speed (V_a), and mean radiant temperature (T_{mrt}) were simulated by the ENVI-met model and extracted at designated points in the computational domain and subsequently imported into the RayMan model to estimate physiological equivalent temperature (PET). It is because PET calculation was not included in the

Table 7.1 Thermal sensation categories for Hong Kong (Cheng et al. 2012; Ng and Cheng 2012)

PET (°C)	Thermal sensation	Physiological stress
<13	Very cold	Extreme cold stress
13–17	Cold	Strong cold stress
17–21	Cool	Moderate cold stress
21–25	Slightly cool	Slightly cold stress
25–29	Neutral	No thermal stress
29–33	Slightly warm	Slightly heat stress
33–37	Warm	Moderate heat stress
37–41	Hot	Strong heat stress
41	Very hot	Extreme heat stress

basic version of ENVI-met used for the present study. The calculated PET was then classified into various categories of thermal sensation (Table 7.1) originated from previous studies of outdoor thermal comfort in Hong Kong (Cheng et al. 2012; Ng and Cheng 2012) where the neutral PET range was found to be 25–29 °C and similar to that of similar climate (Hwang et al. 2011).

7.2.3 Validation of the ENVI-Met Model

Field measurements were conducted on 23rd August as well as on 15th and 17th October 2016 between 09 and 17 h local time in a neighbourhood dominated by commercial and industrial buildings (Fig. 7.2) located in the east of the Kowloon Peninsula. The ENVI-met model was configured for the neighbourhood using:

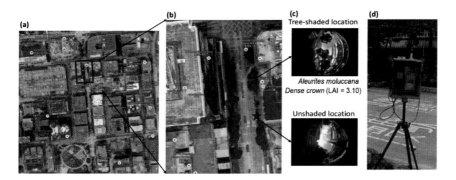

Fig. 7.2 **a** Google map of study area; **b** zoom-in image of street canyon, red stars indicate measurement points; **c** fish-eye images of the measurement locations; and **d** mobile meteorological station

(1) Urban morphology: The building layout and street network, as well as dimension information of the neighbourhood, were obtained from the iB1000 topographic dataset of the Hong Kong Lands Department. The building (including podium) height ranges between 20 and 220 m, and the dominant building materials were assumed to be concrete while the streets are mainly composed of concrete overlaid with asphalt. They are in the NW–SE and E-W orientations and are 15–30 m in width.

(2) Greenery information: A site visit was conducted to obtain information about the number, spacing, height, and width of trees in the simulation domain. In general, the study domain contained several rows of street trees, including dense-crown *Aleurites moluccana* (Candlenut) and sparse-crown *Melaleuca quinquenervia* (paperbark). Hemispherical photographs of the two tree species were imported to Hemisfer software (Thimonier et al. 2010) to estimate the leaf area index. In the present study, the estimated leaf area index (LAI) by Miller's (1967) Li-Cor LAI-2000 algorithm was selected which gives a LAI of 3.10 and 2.20 for *A. moluccana* and *M. quinquenervia*, respectively. As ENVI-met simulates vegetation based on leaf area density (LAD) and not LAI, Eq. (7.1) (Bruse and Fleer 1998; Skelhorn et al. 2014), which shows relationship between the two, was used to estimate the corresponding LAD for each height of the tree given the tree's height, width, and trunk height:

$$
\text{LAI} = \int_0^h \text{LAD}.\Delta z \tag{7.1}
$$

where h is the height of the tree (m) and Δz is vertical grid size (m). LAI and LAD are the leaf area index and leaf area density (m^2/m^3).

(3) Meteorological data: Hourly meteorological variables of each of the measurement days were obtained from the nearest Hong Kong Observatory station and employed as the boundary conditions for the corresponding simulation day. Table 7.2 details the summary of all input parameters and values for the validation simulation.

7.2.4 Field Measurements and Validation Results

In a north–south oriented street canyon in the study area, two mobile meteorological stations were set up at two locations—one under an *A. moluccana* tree (tree-shaded) and another at a nearby open (unshaded) location. The mobile meteorological station consisted of a TESTO480 data logger and sensors at 1.5 m height to simultaneously measure T_a, RH, V_a, globe temperature (T_g) at 10 s sampling interval. The measured data were then resampled into hourly data to match with ENVI-met output temporal resolution. T_{mrt} was calculated by using Eq. (7.2) (Tan et al. 2016):

Table 7.2 Summary of input, test parameters, and corresponding values for validation simulation

Parameter	Definition	Input value
Urban morphology	Street orientation	NW–SE and E-W
	Street width (m)	20–220
	Wall, road, and roof albedo	0.3
Greenery information	Crown width (m)	
	Aleurites moluccana	7
	Melaleuca quinquenervia	3
	Height (m)	
	Aleurites moluccana	10
	Melaleuca quinquenervia	7
	Trunk height (m)	
	Aleurites moluccana	3
	Melaleuca quinquenervia	2
	Leaf area index (m)	
	Aleurites moluccana	3.10
	Melaleuca quinquenervia	2.20
Meteorological data	Initial air temperature (°C)	Hourly profile for each day
	Relative humidity (%)	Hourly profile for each day
	Inflow direction	90° (23rd Aug), 300° (15th and 17th Oct)
	Wind speed at 10 m	2.83 (23rd Aug), 3.0 (15th and 17th Oct)
	Soil temperature (°C)	Default

$$T_{mrt} = \left[\left(T_g + 273.15\right)^4 + \frac{1.1 \times 10^8 V_a^{0.6}}{\varepsilon D^{0.4}} \times \left(T_g - T_a\right) \right]^{0.25} - 273.15 \qquad (7.2)$$

where T_{mrt} is the mean radiant temperature, T_g and T_a refer to globe and air temperature, respectively, V_a is the air speed. D is the diameter of the globe (30 mm) and ε is the emissivity of the globe (0.95).

There was generally a similar pattern with slight discrepancies between observed and simulated hourly T_a and T_{mrt} at both shaded and open locations. Slight reduction in T_a was observed but considerable reduction in T_{mrt} was found, due to the presence of tree. Such a reduction was not owing to the shading by surrounding buildings as both locations were in the street canyon such that the shadow-cast effect was solely due to the tree-shaded effect. However, the ENVI-met model underestimated daytime T_a by 20–30%, which is likely due to the anthropogenic heat sources from human activities, vehicle emission, and mechanical cooling systems not accounted for in the model. In contrast, T_{mrt} was overestimated by 10–40% due to the stable sky conditions assumed in the simulation runs. Other shortcomings of the ENVI-met model for the microclimatic simulation of a built environment include single wall material and the corresponding properties (albedo and the thermal transmittance)

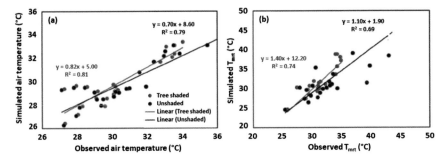

Fig. 7.3 Relationship between ENVI-met simulated and observed **a** T_a and **b** T_{mrt} at tree-shaded and unshaded location on 23rd August, 15th and 17th October 2016

assumed which is not so in reality and static wind conditions (speed and direction) all through the simulation. Nonetheless, fairly good correlations between the measured and simulated values were observed in T_a and T_{mrt} ($R^2 = 0.79$–0.81 and 0.70–0.74, respectively; Fig. 7.3), suggesting that the ENVI-met model is capable of investigating the plant–surface–atmosphere interactions in a complex urban environment. The validation results in the present study corroborate previous studies in Hong Kong (Ng et al. 2012; Tan et al. 2016) which also found a reasonable agreement between measured and simulated microclimatic conditions.

7.2.5 Model Parameterisation

In the present study, simulation runs were performed to represent the subtropical hot–humid summer climate in Hong Kong that is characterised by high-density as well as high-rise urban form and morphology. Hong Kong is located on the south-eastern coast of China (22° 15′ N, 114° 10′ E) with a subtropical climate and a summer average temperature of approximately 28.5 °C. The model forcing technique was used to force hourly T_a and RH of a typical summer day 23rd August 2016 at the model boundary for higher accuracy. Minimum and maximum T_a were 26.2 °C and 33.7 °C, respectively while RH ranged from 60 to 89%. Wind speed at 10 m above ground was set to 3 m/s while the inflow direction was set to 90° to represent a street canyon with limited ventilation. The initial soil temperature was set to 20 °C while the wall, road, and roof albedo were all set to 0.3. Simulation time for each run was 2 days (48 h) while the output of the last 24 h (07–06 h) was analysed to capture the daytime and night-time period (the previous data were taken as spin up). Summary of all input parameters and values for the parametric study can be found in Table 7.3.

Table 7.3 Summary of input, test parameters, and corresponding values for the parametric study

Parameter	Definition	Input value
Meteorological conditions	Initial air temperature (°C)	Hourly profile
	Relative humidity (%)	Hourly profile
	Inflow direction	90°
	Wind speed at 10 m (m/s)	3
	Soil temperature (°C)	20
Street canyon layout	Length (m)	60
	Street width (m)	Refer to Table 7.4
	Building height (m)	Refer to Table 7.4
	Aspect ratio	Refer to Table 7.4
	Wall, road, and roof albedo	0.3
Street trees	Crown diameter (m)	Refer to Table 7.5
	Tree height (m)	
	Trunk height (m)	
	Leaf area index	

Table 7.4 Features and dimensions of street and building layout

Street Type	FP (m)	PL (m)	MR (m)	W (m)	H (m)	AR (H/W)	Case Code
Street layout							
Double carriageway	2	3	20	30	40	1.33	DCW_AR = 1.33
	2	3	20	30	60	2.0	DCW_AR = 2
Single carriageway	2	3	10	20	40	2.0	SCW_AR = 2
	2	3	10	20	60	3.0	SCW_AR = 3

7.2.6 Configuration of Parametric Study

Four street layouts were considered in the parametric study (Fig. 7.4):

1. Open area without trees (OA_{tf}) represents open streets with no nearby buildings and trees.
2. Open area with trees (OA_{wt}) represents open streets with no nearby buildings but with tree shades.
3. Street canyon without trees (ST_{tf}) represents a tree-free street canyon.
4. Street canyon with trees (ST_{wt}) represents street canyons with trees.

The aspect ratio of the street canyons was derived from the average building height and street width of Hong Kong. The building height information of one of the densest districts, Tsim Sha Tsui, where there are over 1400 buildings and the average

Table 7.5 Physical configuration of tree species selected in the present study

Code	Species name	Leaf type	H_T (m)	TH (m)	CH (m)	CW (m)	LAI (m^2/m^3)	TM (%)
T1	*Roystonea regia*	Evergreen	13	9	4	6	1.10	51.6
T2	*Casuarina equisetifolia*	Evergreen	14	4	10	7	1.52	30.3
T3	*Bombax malabaricum*	Deciduous	6	3	3	7	1.83	35.5
T4	*Livistona chinensis*	Evergreen	11	6	5	6	2.11	23.0
T5	*Aleurites moluccana*	Evergreen	9	3	6	7	2.77	18.6
T6	*Macaranga tanarius*	Evergreen	4	1	3	8	3.02	16.2
T7	*Melaleuca leucadendron*	Evergreen	11	3	8	6	3.42	23.5
T8	*Bauhinia blakeana*	Evergreen	7	2	5	6	3.55	10.6

H_T—Height of the tree; TH—Trunk height; CH—Crown height; CW—Crown diameter width; LAI—Leaf area index; TM—Transmissivity of downward radiation (%)

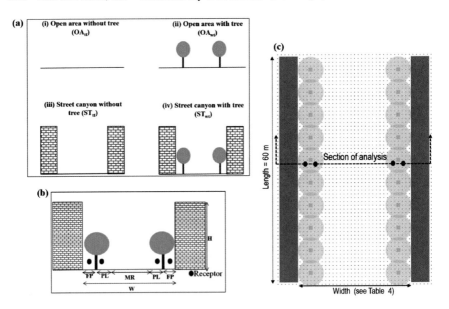

Fig. 7.4 **a** Schematic diagram of four street layout; Generic street layout with embedded features, **b** vertical (XZ) view **c** Plan (XY) view showing the section of analysis [FP: Footpath; PL: Parking Lane; MR: Main road; H: Building height; W: Street width]

building height is 42.5 m with a standard deviation of 27.0 m (Chen et al. 2012). Hence, two building heights, 40 and 60 m, were used in the present study. Two broad categories adopted by Hong Kong Transport Department, namely single carriageway (SCW) and double carriageway (DCW), which is approximately 20 m and 30 m wide, respectively, including footpath and parking lane, were used to determine the typical street width in the present study. It therefore resulted in four aspect ratios with a road length of 60 m, and a symmetrical north–south orientation was adopted in the simulation runs. Table 7.4 details the dimension of road layout and building height while a schematic layout is shown in Fig. 7.4.

Eight tree species were selected to test for improvement of human thermal comfort for the case of open area with trees (OA_{wt}) and street canyon with trees (ST_{wt}). Their selection was based on their popularity of amenity planting in Hong Kong (Zhang and Jim 2014) and their crown width (<8 m) such that they can be planted inside a typical and realistic street canyon of the compact settings in Hong Kong. A representative tree was sampled on the campus of the Chinese University of Hong Kong for field measurement of tree height (H_T), trunk height (TH), crown height (CH), and crown width (CW). Leaf area index (LAI) and transmissivity of solar radiation (TM) of the selected tree species were estimated using Hemisfer software (Thimonier et al. 2010) based on the hemispherical photographs captured on field (Fig. 7.5). For each tree-planted scenario, 20 trees were planted with 10 on each side of the canyon with 6 m spacing (see Fig. 4c for illustration). Table 7.5 lists a summary of the physical and morphological configuration of selected trees.

46 cases were developed, with the street layout, canyon dimension, and tree configuration taken into consideration. The computational domain covered a horizontal area of 40 m × 60 m × 70 m and 30 m × 60 m × 70 m (with 1 m × 1 m × 2 m grid size) for double carriageway (DCW) and single carriageway (SCW) street types, respectively. 10 nested grids were also added to the computational domain in order

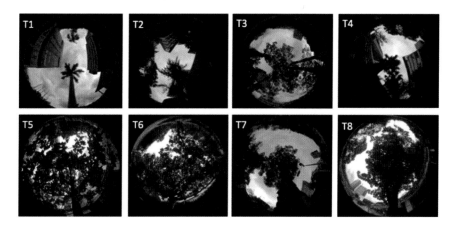

Fig. 7.5 Hemisphere photographs of the eight trees considered in the present study

to ensure sufficient distance before the upstream buildings and after the downstream buildings and minimise edge effect.

Physiological equivalent temperature (PET) was selected for quantifying the conditions of human thermal comfort. Hourly meteorological data required for PET calculation were obtained at four points located in the middle of the street canyon ($y = 30$ m) for uniform and balanced characterisation and comparison. The points were located 2 m and 5 m from both walls to imply footpaths and parking lanes, respectively so that the effect of near-wall computational errors on extracted variables can be minimised and the thermal conditions of where pedestrians mostly use can be captured. The averaged values of meteorological variables from the four points were reported as a single PET value for any cases considered.

The microclimate and thermal comfort within the street canyon can be determined by the shadow and tree-shading effect. Previous studies investigated the effect of trees on human thermal comfort in the street canyons by the difference between PET without and with trees while the dominant shadow-cast also influences thermal conditions. In order to separate the influence of tree shading and shadow-cast and quantify the effect of one over the other, the following diagnostic equations were applied to the simulation results for subsequent analysis.

The impact of only shadow-cast by bounding buildings (of varying height) was estimated as:

$$\Delta PET_{shd} = PET_{sc\_tf} - PET_{oa\_tf} \tag{7.3}$$

where ΔPET_{shd} is the effect of shadow-cast on thermal comfort within the street canyons, as determined by the aspect ratio of the canyons. PET_{sc_tf} and PET_{oa_tf} are the PET in the tree-free street canyon and open area, respectively.

The influence of tree shading (depending on the tree species and its physical configuration) on thermal comfort in open area can then be determined by:

$$\Delta PET_{tree\_oa} = PET_{oa\_wt} - PET_{oa\_tf} \tag{7.4}$$

where ΔPET_{tree_oa} is the effect of tree shading on thermal comfort in open area. PET_{oa_tf} and PET_{oa_wt} are the PET in open area without and with trees, respectively.

The influence of tree shading (depending on the tree species and its physical configuration) on thermal comfort in street canyons can then be determined by:

$$\Delta PET_{tree\_sc} = PET_{sc\_wt} - PET_{sc\_tf} \tag{7.5}$$

where ΔPET_{tree_sc} is the effect of shadow-cast and tree shading on thermal comfort in street canyons. PET_{sc_tf} and PET_{sc_wt} are the PET in street canyons without and with trees, respectively.

The impact of bounding buildings (or shadow-cast) on the thermal comfort performance of trees in street canyons can be estimated as:

$$\Delta PET_{shd\_tree} = PET_{tree\_sc} - PET_{tree\_oa} \qquad (7.6)$$

where ΔPET_{shd_tree} is the effect of shadow-cast on tree performance.

7.3 Results and Discussion

7.3.1 Impacts of Shadow-Cast on Thermal Comfort in Street Canyons

The impact of the shadow of bounding buildings (i.e. ΔPET_{shd} in Eq. (7.3)) on the thermal comfort conditions in street canyons was determined by the difference between PET of a tree-free open area and street canyons with the four densities considered, namely DCW_AR = 1.33, DCW_AR = 2, SCW_AR = 2, and SCW_AR = 3 (Fig. 7.6). Negative ΔPET_{shd} (improvement in thermal comfort) of -0.8 to -12 °C was observed during daytime while positive values of 0.7–2.4 °C were observed at night-time. It is in line with the findings of previous studies that the

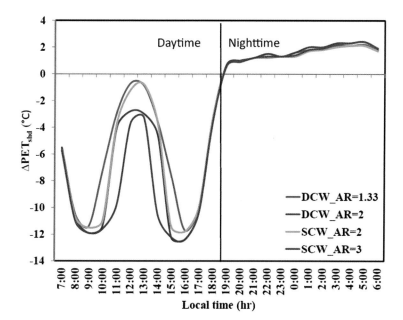

Fig. 7.6 Influence of shadow-cast only on in-canyon thermal comfort during daytime and night-time

shadowing effect reduces surface temperature and radiant energy and hence improves thermal comfort. It is important to note that the effect observed during night-time does not significantly vary across the aspect ratios unlike the daytime observations. The magnitude of ΔPET_{shd} from 11 to 14 h reduced with increasing deepness of the street canyon, with the exception found between the difference between DCW_AR $= 2$ and SCW_AR $= 2$. The latter indicates higher negative ΔPET_{shd} values than the former from 11 to 12 h in spite of similar aspect ratios. This suggests that the deepness or shallowness of canyon is more responsive than the aspect ratio value in terms of the thermal comfort improvement by shadow-cast. Although the ventilation is reduced by 90–96% by the bounding buildings compared to the open area especially under perpendicular wind direction, the reduction in PET was more influenced by the low T_{mrt} caused by the shadow of the buildings.

7.3.2   Effect of Tree Species on Thermal Comfort in Open Area

Figure 7.7 shows the temporal variation of percentage solar attenuation by each tree species. Data of 12 and 15 h were selected for analysis as the pattern of other hours is rather similar. At 12 h, *Roystonea regia* (T1) with the least LAI reduces approximately 45% of the incoming solar radiation while *Bauhinia blakeana* (T8) with the highest

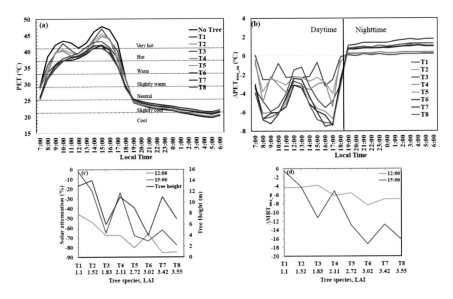

Fig. 7.7   a Temporal variation of PET without and with tree of different species in open area; **b** Hourly variation of (ΔPET_{tree_oa}); **c** Solar attenuation by the tree species at 12 and 15 h; and **d** $\Delta T_{mrt\ tree\_oa}$ by the tree species at 12 and 15 h

LAI reduces approximately 85%. At 15 h, T1 and T8 attenuated approximately 2% and 80%, respectively. It indicates higher potential in solar attenuation at high solar attitude irrespective of tree species. A direct relationship between LAI and solar attenuation was also observed with some exceptions: T4 attenuated less than *Bombax malabaricum* (T3) and *Melaleuca leucadendron* (T7) attenuated less than T6, especially at 15 h even though the later have higher LAI while the former is shorter than the latter (Fig. 7.7c). This suggests that tree height also is another factor accounting for the magnitude of solar attenuation. This result is consistent with a previous study (Fahmy et al. 2010) that *Ficus elastica* with LAI of 3 attenuated 84% of direct solar radiation while *Mesua ferrea* with LAI value of 6.1 and canopy transmissivity of approximately 5% could reduce incoming solar radiation by 93%, thus significantly contributing to the cooling benefit (Shahidan et al. 2010). It also influences the pattern of T_{mrt}, which is the parameter most affected by tree shade, and its dependence on radiative fluxes (especially direct shortwave radiation). For instance, T_{mrt} at 15 h was reduced by 0.5 °C, 4.1 °C, 11.2 °C, 5.1 °C, 12.9 °C, 17.2 °C, 12.7 °C, and 16.2 °C with T1, T2, T3, T4, T5, T6, T7, and T8, respectively. Figure 7.7a describes the hourly variation of PET in open area without and with different tree species. In the tree-free open area, PET was found to be above 41 °C between 09 and 17 h, indicating the corresponding thermal sensation as "very hot". The magnitude of PET was reduced to a "hot" thermal sensation with trees of higher LAI while it remained in a "very hot" conditions with other tree species. The thermal comfort conditions mostly became "slightly cool" or "cool" at night-time regardless of the presence of tree species due to the absence of solar radiation.

The effect of tree species on human thermal comfort in open area (ΔPET_{tree_oa}) during daytime and night-time is shown in Fig. 7.7b. The magnitude of improvement during daytime (15 h) increases with LAI except for *Livistona chinensis* (T4) and *M. leucadendron* (T7). ΔPET_{tree_oa} of *R. regia* (T1) with the least LAI is approximately −1 °C irrespective of the width of the street canyons while *B. blakeana* (T8) with the highest LAI of 3.55 resulted in ΔPET_{tree_oa} of approximately −7 °C. This is apparently due to the solar attenuation which is dependent on LAI and consequently lower Tmrt values. The counteracting process of longwave radiation released during night-time (03 h) caused a positive ΔPET_{tree_oa}. The magnitude of positive ΔPET_{tree_oa} somewhat reduced with increasing LAI due to the trapping of longwave radiation by tree canopy (Ali-Toudert and Mayer 2006; Theeuwes et al. 2014).

7.3.3 Effect of Tree Species on Thermal Comfort in Street Canyons

The effect of different tree species on human thermal comfort at 12 and 15 h is shown in Fig. 7.8. The pattern of cases of street canyons resembles that of the open area except for the magnitude of attenuation. LAI and tree (as well as trunk) height are the major factors for the percentage solar attenuation and hence the $\Delta T_{mrt\ tree\_sc}$. With the

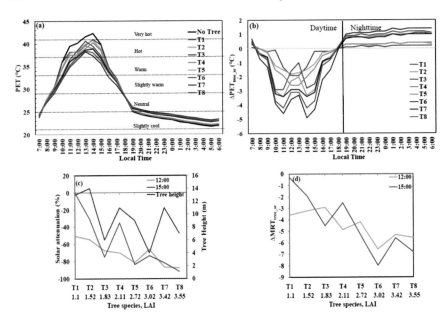

Fig. 7.8 **a** Temporal variation of PET without and with tree of different species in street canyons; **b** hourly variation of (ΔPET_{tree_sc}); **c** solar attenuation by the tree species at 12 and 15 h; and **d** $\Delta T_{mrt\ tree\_sc}$ by the tree species at 12 and 15 h

presence of trees, $\Delta T_{mrt\ tree\_sc}$ at 12 h was reduced by 3.6 °C, 3.2 °C, 2.9 °C, 4.9 °C, 4.2 °C, 6.6 °C, and 5.3 °C for T1, T2, T3, T4, T5, T6, T7, and T8, respectively.

The effect of tree species on the temporal variation of PET in street canyons is shown in Fig. 8a, with the case DCW_AR = 1.33 shown for illustration. Only one peak in PET was observed in street canyons when the sun was overhead, compared to the two peaks observed in open area. Without the presence of trees, PET was found to exceed 41 °C between 12 and 15 h, indicating a "very hot" thermal condition. It was reduced with the presence of trees with the attenuation varying across tree species. From 19 h onwards, the thermal conditions greatly improved to "neutral" or "slightly cool". In general, all tree species led to improved and worsened thermal comfort conditions (ΔPET_{tree_sc}) during daytime and night-time respectively (Fig. 8b). Figure 7.9 shows that the magnitude of ΔPET_{tree_sc} varied with species and the aspect ratio of the street canyons. The magnitude of negative ΔPET_{tree_sc} was higher in DCW_AR = 1.33, with slight variations in other street canyons regardless of tree species. This reiterates the need for tree planting to improve human thermal comfort in pedestrian environment decreases with the deepness of street canyons during daytime due to the overwhelming shadowing effect. The influence of the aspect ratio of street canyons diminished during night-time. Similar to the observation in the open area, negative ΔPET_{tree_sc} during daytime increased with the LAI of trees except for *L. chinensis* (T4) and *M. Leucadendron* (T7) which are equally responsive to tree height. This pattern was also observed in solar attenuation and

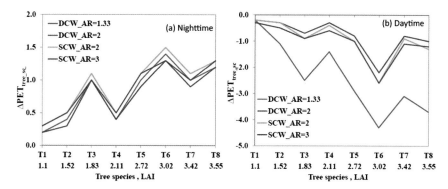

Fig. 7.9 Impact of trees on in-canyon thermal comfort at **a** night-time (03 h) and **b** daytime (15 h)

reduction in T_{mrt}. Due to the solar attenuation, the amount of upward longwave radiation trapped and hence positive ΔPET_{tree_sc} also increased with LAI except for T4 and T7.

7.3.4 Impact of Shadow-Cast on In-Canyon Tree Potential

The shadow-cast effect of bounding buildings on the cooling potential of tree species for human thermal comfort during daytime is shown in Fig. 7.10. Positive ΔPET_{shd_tree} values show that bounding buildings reduced the potential in PET reduction and vice versa for negative values. The performance of all tree species declined when they were planted in street canyons. The magnitude somewhat varied

Fig. 7.10 Impact of shadow-cast on in-canyon tree potential for thermal comfort improvement at 15 h

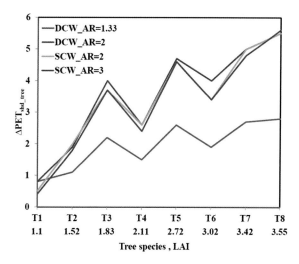

with tree species and the aspect ratio of street canyons. For example, *R. regia* (T1) reduced PET by 1 °C when planted in open area while it reduced PET only by 0.2 °C when planted in street canyon (DCW_AR = 1.33), suggesting that the bounding buildings have a negative impact on the performance of trees. *B. blakeana* (T8), with the highest LAI, reduced PET at open area by 6.5 °C but it was reduced to 3.7 °C when planted in street canyon (DCW_AR = 1.33). Generally speaking, the potential of tree in improving human thermal comfort reduces in street canyons due to the restricted ventilation and lowered T_{mrt}. This effect is more prominent when the LAI of trees increases and/or when the street canton becomes deeper.

7.3.5 Relationship Between Trees' Physical Configuration and Thermal Effect

A regression analysis was conducted to obtain a holistic understanding of how each parameter of the tree's physical configuration contributes to thermal comfort in different urban densities during daytime and night-time. Table 7.6 shows the correlation coefficient (R^2) of each parameter and $\Delta PET_{tree\_sc}/\Delta PET_{tree\_oa}$ for daytime and night-time, which were classified into level of importance and detriment for daytime and night-time, respectively. LAI and TH were found to be the "most important" parameters for human thermal comfort in both open area and shallow street canyons (DCW_AR = 1.33) during daytime, indicating that high LAI and low TH lead to larger reduction in PET due to higher solar attenuation as the magnitude of LAI and TH increases and reduce, respectively. Due to larger impact of shadow-cast in deeper street canyons, the effect of LAI and TH was reduced to "less important" and "more important", respectively, while H_T maintains as "more important". In contrast, regardless of the canyon density, CH and CW were "less" and "least" important to human thermal comfort during daytime. It therefore reiterates that tall trees with low LAI and high trunk in deep street canyons and vice versa for more open settings (open area and shallow street canyons) have more a prominent effect on human thermal comfort during daytime.

During night-time, the trapping of longwave radiation leads to nocturnal urban heat islands and deteriorating levels of thermal comfort. From the simulation results, H_T and TH, followed by LAI, were the most influential parameters of tree configuration resulting in such an effect. It implies that shorter trees and trunk height with higher LAI lead to higher-temperature nocturnal urban heat islands and hence thermal discomfort in the urban areas irrespective of urban density. This study revealed an inverse effect on the impact of a tree's configuration, particularly LAI, TH, and H_T, on the thermal environment during daytime and night-time.

7.4 Recommended Street Tree-Planting Strategies

With the understanding obtained from this parametric study, the following recommendations are formulated for urban designers and landscape architects in order to maximise the multiple benefits of trees in the dense urban environment.

(a) Tree planting should be implemented irrespective of urban geometry
Our results show that the need and effectiveness of tree planting in improving human thermal comfort in street canyons are reduced with the deepness of street canyons due to the prominent shadowing effect. Nevertheless, urban designers and landscape architects should consider tree planting for a wide range of ecosystem services such as aesthetic benefits, energy saving, storm water reduction, pollutant removal, and carbon dioxide sequestration (Morani et al. 2011; Wang et al. 2014; Mullaney et al. 2015).

(b) Less canopy density in deeper canyon and vice versa
Since the shadowing effect is more effective in deeper canyons and vice versa for shallower canyons, the selection of tree species should depend on the urban morphology or geometry. Hence, we propose tall and lower LAI trees such as *Casuarina equisetifolia* (T2), *B. malabaricum* (T3), and others with similar configuration for deep street canyons in Hong Kong. This configuration will improve human thermal comfort during both daytime and night-time while the dispersion of air pollutants and local ventilation will be less hampered. In contrast, high LAI trees with medium height such as *B. blakeana* (T5), *Macaranga tanarius* (T6), and *A. moluccana* (T8) and other similar trees species (including those with larger crown width) are recommended for tree planting in shallow street canyons and open spaces such as urban parks, car parks, and other recreational centres where the shadowing effect of bounding buildings is limited.

(c) Higher trunk height for deeper canyons and vice versa
The results of this study show that trunk height is influential on the human thermal comfort during daytime and night-time. However, trunk height is not a determining factor if the LAI of trees is similar across different species. Higher trunk height allows for improved ventilation, and pollutant dispersion is more important in deep street canyons. It is therefore recommended that trunk base of the trees should be maintained to increase with the deepness of street canyons. This provides additional vertically accessible spaces under tree canopies for pedestrians and duty vehicles. Based on the study findings, a minimum trunk height of 2 m is suggested in open areas and shallow street canyons while such trunk height should be doubled in deeper street canyons to maintain the benefits. However, this condition should not be compromised for canopy density in any case.

7.5 Conclusions and Further Work

Urban trees provide multiple ecosystem service and therefore should be appropriately incorporated into the planning and implementation phases of urban development. In this study, an ENVI-met model was employed to quantify the corresponding contribution of shadow-cast and tree shading to the improvement of human thermal comfort in the high-density urban environment of Hong Kong. In particular, eight common tree species were examined for their effect on pedestrian thermal comfort in summer since their microclimatic benefits are highly dependent on their physical properties.

Our results show that the leaf area index is the most influential factor to thermal benefits of urban trees during daytime although its influence during night-time is considerably reduced. Nonetheless, the contribution of other parameters of the physical configuration such as tree height and trunk height cannot be overemphasised. Our study also revealed that the potential of trees for improving thermal comfort reduces with increasing urban density due to the shadowing effect of surrounding buildings. However, tree planting should be encouraged irrespective of the urban density as trees provide a wide range of ecosystem services. Based on our results, tall trees with higher trunks and short crown widths are proposed for tree planting in high-density areas where the shadowing effect is dominated while the reverse is proposed for more open areas.

Beyond the present study, a combination of various strategies should be adopted in designing pedestrian streetscapes. For instance, colonnades, overhanging façade, and galleries are possible means to supplement the shading of street trees in subtropical or tropical regions with high summer solar altitude (Johansson 2006; Ali-Toudert and Mayer 2007). Water ponds and fountains can also be considered in open spaces to provide additional evaporative cooling but should not be placed in high-density areas where air movement is limited (Theeuwes et al. 2013; Steeneveld et al. 2014). Other techniques such as improved canyon ventilation and reflective surface materials may be applied in urban design especially in tropical climates (Lobaccaro and Acero 2015; Morakinyo and Lam 2016). These mitigation measures should be collectively considered to reduce the thermal load of the urban environment.

There are some limitations of the ENVI-met model adopted in the present study. Firstly, the walls of bounding buildings in the street canyons had similar thermal behaviour since single wall material was assumed in the model, which may not always be true in reality. Results presented in this study are relevant for symmetrical urban canyons in tropical cities with hot–humid climate. Moreover, we only considered perpendicular inflow condition (implying a canyon with limited ventilation), with other wind direction and higher wind speed the thermal conditions may be improved. In future work, the considered tree species and more could be planted in asymmetrical street canyons and other street orientations which are more typical of real urban canyons for more comprehensive conclusions. Future studies could also consider the economic factors of candidate trees beyond the environmental benefits to ensure holistic consideration for choosing tree species for urban greenery projects.

In addition, some of the less common species not considered in this study could even have better thermal benefits and therefore deserve more research attention. Also, the spread of trees species depends on geographical location. As such, the species proposed for implementation in this study may be applicable only to Hong Kong and some other cities. Nonetheless, the different species with similar configuration and properties may yield similar or better results. Furthermore, we propose the integration of the trees' growth rates into our consideration of tree selection for urban planting. Fast-growing species should be considered for quicker utilisation of their cooling and other benefits. However, an improved and frequent maintenance culture must be imbibed as trees' growth rate and size may contest with limited spaces especially in high-density area. Future research in plant science and engineering could be geared towards the development of climate-sensitive fast-growing trees.

References

Ali-Toudert, F., and H. Mayer. 2006. Numerical study on the effects of aspect ratio and orientation of an urban street canyon on outdoor thermal comfort in hot and dry climate. *Building and Environment* 41: 94–108.

Ali-Toudert, F., and H. Mayer. 2007. Effects of asymmetry, galleries, overhanging façades and vegetation on thermal comfort in urban street canyons. *Solar Energy* 81: 742–754.

Armson, D., P. Stringer, and A.R. Ennos. 2012. The effect of tree shade and grass on surface and globe temperatures in an urban area. *Urban Forestry & Urban Greening* 11: 245–255.

Armson, D., M.A. Rahman, and A.R. Ennos. 2013. A comparison of the shading effectiveness of five different street tree species in Manchester, UK. *Arboriculture and Urban Forestry* 39: 157–164.

Berry, R., S.J. Livesley, and L. Aye. 2013. Tree canopy shade impacts on solar irradiance received by building walls and their surface temperature. *Building and Environment* 69: 91–100.

Bowler, D.E., L. Buyung-Ali, T.M. Knight, and A.S. Pullin. 2010. Urban greening to cool towns and cities: A systematic review of the empirical evidence. *Landscape and Urban Planning* 97: 147–155.

Bruse, M., and H. Fleer. 1998. Simulating surface-plant-air interactions inside urban environments with a three-dimensional numerical model. *Environmental Modelling and Software* 13: 373–384.

Chen, L., E. Ng, X. An, C. Ren, M. Lee, U. Wang, and Z. He. 2012. Sky view factor analysis of street canyons and its implications for daytime intra-urban air temperature differentials in high-rise, high-density urban areas of Hong Kong: A GIS-based simulation approach. *International Journal of Climatology* 32: 121–136.

Cheng, V., E. Ng, C. Chan, and B. Givoni. 2012. Outdoor thermal comfort study in a subtropical climate: A longitudinal study based in Hong Kong. *International Journal of Biometeorology* 56: 43–56.

de Abreu-Harbich, L.V., L.C. Labaki, and A. Matzarakis. 2015. Effect of tree planting design and tree species on human thermal comfort in the tropics. *Landscape and Urban Planning* 138: 99–109.

Fahmy, M., S. Sharples, and M. Yahiya. 2010. LAI based trees selection for mid latitude urban developments: A microclimatic study in Cairo, Egypt. *Building and Environment* 45: 345–357.

Steeneveld, G.J., S. Koopmans, B.G. Heusinkveld, and N.E. Theeuwes. 2014. Refreshing the role of open water surfaces on mitigating the maximum urban heat island effect. *Landscape and Urban Planning* 121: 92–96.

Gómez-Baggethun, E., and D.N. Barton. 2013. Classifying and valuing ecosystem services for urban planning. *Ecological Economics* 86: 235–245.

Huttner, S., M. Bruse, P. Dostal, A. Katzschner, and J. Gutenberg-universität. 2009. Strategies for mitigating thermal heat stress in central European cities: The project klimes. In *The 7th International Conference on Urban Climate* 49: 3927089, 29 June–3 July 2009.

Hwang, R.L., T.P. Lin, and A. Matzarakis. 2011. Seasonal effects of urban street shading on long-term outdoor thermal comfort. *Building and Environment* 46: 863–870.

Johansson, E. 2006. Influence of urban geometry on outdoor thermal comfort in a hot dry climate: A study in Fez, Morocco. *Building and Environment* 41: 1326–1338.

Lobaccaro, G., and J.A. Acero. 2015. Comparative analysis of green actions to improve outdoor thermal comfort inside typical urban street canyons. *Urban Climate* 14: 251–267.

Matzarakis, A., F. Rutz, and H. Mayer. 2010. Modelling radiation fluxes in simple and complex environments: Basics of the RayMan model. *International Journal of Biometeorology* 54: 131–139.

McPherson, E.G., J.R. Simpson, Q. Xiao, and C. Wu. 2011. Million trees Los Angeles canopy cover and benefit assessment. *Landscape and Urban Planning* 99 (1): 40–50.

Miller, J.B. 1967. A formula for average foliage density. *Australian Journal of Botany* 15 (1): 141–144.

Morakinyo, T.E., and Y.F. Lam. 2016. Simulation study on the impact of tree configuration, planting pattern and wind condition on street-canyon's micro-climate and thermal comfort. *Building and Environment* 103: 262–275.

Morani, A., D.J. Nowak, S. Hirabayashi, and C. Calfapietra. 2011. How to select the best tree planting locations to enhance air pollution removal in the Million-Trees NYC initiative. *Environmental Pollution* 159: 1040–1047.

Mullaney, J., T. Lucke, and S.J. Trueman. 2015. A review of benefits and challenges in growing street trees in paved urban environments. *Landscape and Urban Planning* 134: 157–166.

Ng, E., and V. Cheng. 2012. Urban human thermal comfort in hot and humid Hong Kong. *Energy and Buildings* 55: 51–65.

Ng, E., L. Chen, Y. Wang, and C. Yuan. 2012. A study on the cooling effects of greening in a high-density city: An experience from Hong Kong. *Building and Environment* 47: 256–271.

Oke, T.R., J.M. Crowther, K.G. McNaughton, J.L. Monteith, and B. Gardiner. 1989. The micrometeorology of the urban forest. *Philosophical Transactions—Royal Society of London, B* 324: 335–349.

Qaid, A., and D.R. Ossen. 2014. Effect of asymmetrical street aspect ratios on microclimates in hot, humid regions. *International Journal of Biometeorology* 59: 657–677.

Salata, F., I. Golasi, Vollaro R. de Lieto, and Vollaro A. de Lieto. 2016. Urban microclimate and outdoor thermal comfort. A proper procedure to fit ENVI-met simulation outputs to experimental data. *Sustainable Cities and Society* 26: 318–343.

Shahidan, M.F., M.K.M. Shariff, P. Jones, E. Salleh, and A.M. Abdullah. 2010. A comparison of Mesua ferrea L. and Hura crepitans L. for shade creation and radiation modification in improving thermal comfort. *Landscape and Urban Planning* 97: 168–181.

Skelhorn, C., S. Lindley, and G. Levermore. 2014. The impact of vegetation types on air and surface temperatures in a temperate city: A fine scale assessment in Manchester, UK. *Landscape and Urban Planning* 121: 129–140.

Tan, C.L., N.H. Wong, and S.K. Jusuf. 2013. Outdoor mean radiant temperature estimation in the tropical urban environment. *Building and Environment* 64: 118–129.

Tan, Z., K.K.L. Lau, and E. Ng. 2016. Urban tree design approaches for mitigating daytime urban heat island effects in a high-density urban environment. *Energy and Buildings* 114: 265–274.

Theeuwes, N.E., A. Solcerova, and G.J. Steeneveld. 2013. Modeling the influence of open water surfaces on the summertime temperature and thermal comfort in the city. *Journal of Geophysical Research: Atmospheres* 118: 8881–8896.

Theeuwes, N.E., G.J. Steeneveld, R.J. Ronda, B.G. Heusinkveld, L.W.A. van Hove, and A.A.M. Holtslag. 2014. Seasonal dependence of the urban heat island on the street canyon aspect ratio. *Quarterly Journal of the Royal Meteorological Society* 140 (684): 2197–2210.

Thimonier, A., I. Sedivy, and P. Schleppi. 2010. Estimating leaf area index in different types of mature forest stands in Switzerland: A comparison of methods. *European Journal of Forest Research* 129: 543–562.

Wang, Y., F. Bakker, R. de Groot, and H. Wortche. 2014. Effect of ecosystem services provided by urban green infrastructure on indoor environment: A literature review. *Building and Environment* 77: 88–100.

Zhang, H., and C.Y. Jim. 2014. Contributions of landscape trees in public housing estates to urban biodiversity in Hong Kong. *Urban Forestry & Urban Greening* 13: 272–284.

Part III
Applications of Human Thermal Comfort in Urban Planning and Design

Chapter 8
Urban Climatic Map: Thermal Comfort as the Synergising Indicator

Abstract The distinctive features of urban climate have been widely studied for decades. However, the consideration of urban climate in urban planning and design framework is rather limited. One of the possible reasons is due to the difference in working languages between scientists and urban planners such that the urban climatic knowledge cannot be readily utilised in the planning and design practices. The concept of urban climatic map provides an information platform for the presentation of urban climatic phenomena on a two-dimensional spatial map in a format that can be readily interpreted by urban planners. Areas with environmental problems or sensitive to urban climate can therefore be easily identified for mitigation measures. In this paper, the methodology and results of the urban climatic analysis map, as the first part of the urban climatic map study of Hong Kong, are presented and the implications on urban planning and design practices are discussed.

Keywords Urban climate · Urban planning · Urban climatic map · High-density cities · Subtropical

8.1 Introduction

Rapid urbanisation results in the transformation of surface characteristics of the landscape and human activities in urban areas cause dramatic changes in climatic conditions in urban areas (Oke 1987; Mills 1997). A wide range of urban climatic studies were conducted in the last few decades (Arnfield 2003). However, the incorporation of urban climatic knowledge into urban planning and design frameworks is still limited. This leads to inefficient city planning and building design with regard to urban climate and hence impacts the energy consumption of the city (Bitan 1988).

Urban planners are generally found to have difficulties in understanding scientific findings of urban climatic studies and how these studies are usually constructed on the basis of meteorology, climatology, and atmospheric physics (Eliasson 2000; Mills 2006; Ren et al. 2013). It is therefore necessary to bridge the gap between scientific understandings of urban climatology and urban planning and design practices in order to translate climatic knowledge into planning language (de Schiller and Evans

1990–1991; Alcoforado et al. 2009; Ren et al. 2011). A platform for information exchange is required to assemble climatic information and present the useful and necessary information to different stakeholders (Mills et al. 2010).

An urban climatic map (UCMap) is an information and evaluation tool which integrates urban climatic factors and town planning considerations by presenting urban climatic phenomena on a two-dimensional spatial map in a format that can be readily interpreted by urban planners (Ren et al. 2011, 2013). It is produced by collating various meteorological, topographical, planning, and land use information and "their interrelationship and effects on urban climate are analysed and evaluated spatially and quantitatively" (Ren et al. 2013, p. 2). Various "climatopes" are then defined as the spatial units of a UCMap in order to present areas with similar urban climatic conditions and features (Scherer et al. 1999; Alcoforado et al. 2009). They can also show the spatial variations of urban climatic characteristics and their significance. Urban planners can therefore easily identify areas with environmental problems or sensitive to urban climate and planning recommendations in corresponding areas can assist them to take appropriate actions.

The concept of UCMap was developed in Germany in the 1970s (Matzarakis 2005), and it has been applied in Europe, Asia, and South America (Scherer et al. 1999; Alcoforado et al. 2009). Most of the UCMap studies in the world focus on low-density urban planning and design except for the Thermal Environment Map for Tokyo (Tokyo Metropolitan Government 2005). However, it focuses on the thermal aspect of urban climate only and prevailing wind information and localised air movements are critical to outdoor thermal comfort of a coastal city. For high-density subtropical cities like Hong Kong, an improved methodology is required to accurately delineate the climatopes of the city.

This paper presents the methodology of the development of UCMap of Hong Kong. The UCMap system consists of two major components: the urban climatic analysis map (UC-AnMap) and the urban climatic recommendation map (UC-ReMap). The UC-AnMap, which collates meteorological, topographic, planning, and land use information and analyse their interrelationships and effects on the wind and thermal environment, is documented in this chapter. In this chapter, human thermal comfort in the outdoor environment, as represented by physiological equivalent temperature, is employed as the synergising factor to collate the layers of the UCMap. User survey of outdoor thermal comfort provides necessary information to determine the neutral conditions for typical summer in Hong Kong.

8.2 Methodology

8.2.1 Study Area

Hong Kong is situated at the south-eastern coast of China, and its downtown areas are located on both sides of the Victoria Harbour. A number of high-density residential districts are also found in different parts of the city. Although a large proportion of the territory is covered by natural landscape, greenery is somewhat limited in dense urban areas. The highly urbanised downtown areas result in severe urban heat island (UHI) phenomenon with about 4 °C (Wu et al. 2008). At the beginning of the twentieth century, Hong Kong was dominated by agriculture and fisheries with limited trade activities. Urban development was at a steady rate until the outbreak of the Second World War which caused enormous impacts on the economic activities in Hong Kong. After the Second World War, entrepot trade contributed to the prosperity of Hong Kong economy in the 1960s and the city was subsequently transformed into an industrial city in the following decades. With the intensification of urban development in the last few decades, Hong Kong has transformed into a megacity characterised by high-rise buildings and compacted urban form.

8.2.2 Meteorological Data

Hourly T_a and wind data were acquired from the meteorological stations operated by the Hong Kong Observatory (HKO; Fig. 8.1). The data period of all meteorological stations ends at the end of 2004 but varies from station to station due to the difference in establishment date. The HKO Headquarter has the longest meteorological record of 120 years while most of the stations have a record of only about 10 years. The highest T_a is observed from June to August and thermal heat load due to buildings and anthropogenic activities further increase T_a. Therefore, the summer months are

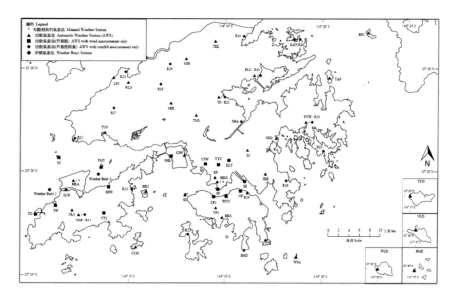

Fig. 8.1 Location of the ground-level meteorological stations operated by HKO

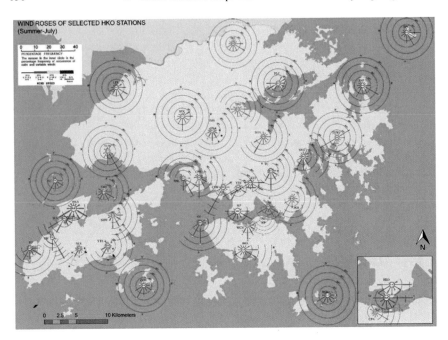

Fig. 8.2 Wind roses of ground-level meteorological stations

generally considered more critical to outdoor thermal comfort in Hong Kong (Ng and Cheng 2012) and the UC-AnMap is developed based on the summer conditions.

Wind data were expertly evaluated in order to provide information on prevailing wind directions under both annual and summer conditions. It helps to gain insights into land and sea breezes, thermal air mass movements, downhill air movements, and topographical effects (Fig. 8.2). The evaluation identifies wind flow patterns such as air paths and air mass exchange, and how they are influenced by prevailing wind was also assessed. In addition, the local conditions of the meteorological were characterised so that background wind, thermally induced circulation systems, and channelling effects were identified. For a more comprehensive understanding of prevailing winds in Hong Kong, the Fifth Generation Mesoscale Model (MM5) was employed to simulate wind speed and direction in Hong Kong for each month in 2004. The grid size is 100×100 m, and the lowest height is 60 m. The MM5 model was previously evaluated with reasonable accuracy (Yim et al. 2007). The 16-direction MM5 wind roses were compared with HKO wind roses for selected locations, and the prevailing wind directions were subsequently determined.

8.2.3 Land, Building, and Planning Data

Land, building, and planning data were acquired from the Planning Department in order to provide information about the topography, urban morphology, and greenery of Hong Kong. The topography of Hong Kong is represented by the digital elevation model (DEM) with a resolution of 2 m. Building and podium data describe the urban morphology in terms of building height and coverage as well as geometrical information. Land use data were used to determine the extent of greenery areas in Hong Kong. In addition, a Normalised Difference Vegetation Index (NDVI) image was also used to supplement the greenery data provided by Planning Department (Nichol and Lee 2005).

8.2.4 Structure of the UC-AnMap

The UC-AnMap adopts the GIS platform which allows urban planners to easily assess and utilise the data. The procedures of developing the UC-AnMap are outlined in Fig. 8.3. The acquired data were used to develop the six layers, namely building volume, topography, green space, ground coverage, natural landscape, and proximity

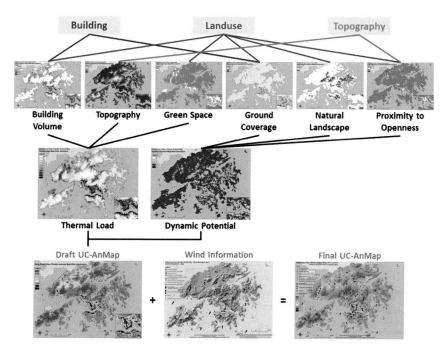

Fig. 8.3 Workflow of the UC-AnMap of Hong Kong

to openness. The layers were then classified into two categories so that the two major components of the draft UC-AnMap, thermal load, and dynamic potential, can be established. With the wind information layer incorporated into the draft UC-AnMap, the final version of the UC-AnMap was obtained.

8.3   User Questionnaire Survey for Outdoor Thermal Comfort

8.3.1   Synergising Indicator: Physiological Equivalent Temperature (PET)

The UCMap of Hong Kong employs physiological equivalent temperature (PET) as a synergising indicator to collate the UC-AnMap since it focuses on outdoor thermal comfort under typical summer conditions of Hong Kong. PET is widely used as a human thermal comfort indicator which is based on the Munich Energy-Balance Model for individuals (MEMI; Fig. 8.4). It is calculated from various climatic and physiological parameters including T_a, relative humidity, solar radiation, air movement, clothing, and metabolic rate in order to give a synergising indication of human thermal comfort in the present study.

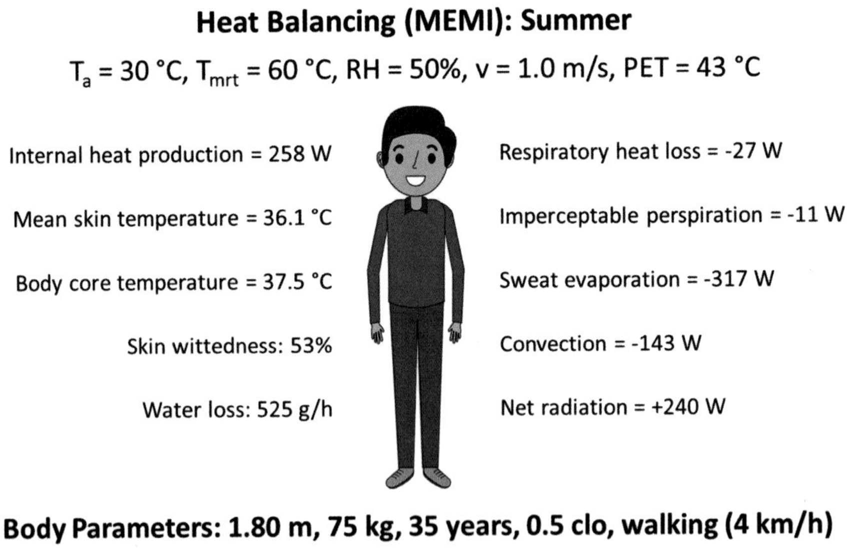

Heat Balancing (MEMI): Summer

$T_a = 30\ °C$, $T_{mrt} = 60\ °C$, RH = 50%, v = 1.0 m/s, PET = 43 °C

Internal heat production = 258 W

Mean skin temperature = 36.1 °C

Body core temperature = 37.5 °C

Skin wittedness: 53%

Water loss: 525 g/h

Respiratory heat loss = -27 W

Imperceptable perspiration = -11 W

Sweat evaporation = -317 W

Convection = -143 W

Net radiation = +240 W

Body Parameters: 1.80 m, 75 kg, 35 years, 0.5 clo, walking (4 km/h)

Fig. 8.4 Munich Energy-Balance Model for individuals (MEMI)

8.3.2 Survey Campaign

The objective of the user questionnaire survey is to obtain the subjective perception of outdoor thermal comfort of local citizens in urban areas of Hong Kong. The sites selected in the present study cover a wide range of microclimatic and environmental conditions, based on parameters related to the regional climatic conditions, topographic characteristics, and urban morphology. The parameters considered in the present study include wind environment, ground coverage, street pattern, building height, and building density. Different microclimatic conditions within a survey site were also considered so that the data obtained from the survey include a wide range of microclimatic conditions people encountered in the urban environment of Hong Kong. The selection criteria of survey sites also included different kinds of land uses and activities, including three types of sites, namely pedestrian streets, residential estates, urban parks. These categories take into account the differences in the nature of activities in the survey sites and the psychological expectation of the people within. A total of 2702 responses were collected from September 2006 to August 2007.

The survey campaign consisted of two parts, namely micrometeorological measurements and a questionnaire survey. Mobile meteorological stations were used to obtain the micrometeorological conditions in the surroundings of the respondents. Each mobile station consisted of sensors for the measurement of dry-bulb air temperature, globe temperature, wind speed, and relative humidity. Humidity ratio was derived by air temperature and relative humidity measurements in the analysis. These meteorological parameters were regarded as the main determinants of outdoor thermal comfort (Penwarden 1973; Gagge and Gonzalez 1974; Tacken 1989; Nagara 1996; Sasaki et al. 2000; Lindberg 2004). A TESTO three-function probe was used to measure air temperature, relative humidity, and wind speed. The temperature and humidity sensors were protected from direct sun exposure by a circular white disc made with polystyrene. Globe temperature was measured using a tailor-made globe thermometer consisting of a thermocouple wire (TESTO flexible Teflon Type K) held at the middle of a 38 mm diameter black table tennis ball (Humphreys 1977; Nikolopoulou et al. 1999). These sensors were connected to a TESTO 400 data logger with sampling and logging time of 5 s.

The questionnaire survey consisted of questions addressing the subjects' thermal comfort condition (e.g. thermal sensation and comfort vote) and also record of subjects' demographic background (gender and age), and clothing and activities during the survey. The subjects' thermal sensation and comfort vote were obtained by face-to-face interviews while the subjects' demographic background, clothing, and activities were observed and recorded by the interviewer conducting the surveys. The results of the questionnaire surveys were subsequently correlated with the micrometeorological data to analyse the general thermal comfort conditions in the outdoor spaces and the comfort requirement of the people in Hong Kong. Table 8.1 details the questions asked in the questionnaire survey. The clothing of the subject was recorded by the interviewer using the garment checklist provided in the questionnaire. The checklist was extracted from ASHRAE Standard 55-2010 and ISO Standard 7730

Table 8.1 Questions and responses of the questionnaire survey

	Questions and responses
1	Have you done this survey before? Yes/No (if the answer is "Yes", the survey will be terminated.) • This is to ensure the subject is not repeatedly sampled
2	Have you been staying in Hong Kong in the past 6 months? Yes/No • To understand the subject's mid- and long-term acclimatisation
3	In the past 15 min prior to the survey, have you been to (or stayed in) air-conditioned indoor spaces (cooled or heated spaces including bus, taxi, minibus, etc.)? Yes/No • To understand the subject's immediate past thermal experience
4	What were you doing in the past 15 min prior to the survey? Waiting for people or cars/resting/standing/sitting/working/grocery/shopping/shopping/doing exercises/others • To understand the subject's immediate past activity
5	Why do you choose to sit/stand at this particular place? (can choose more than one item) In shade/under tree cover/under sunshine/breezy/fresh air/views/have an appointment/no particular reason/going to school or work/close to home or office or school or station/others • The reason for the subject to be at the place of the survey may influence their thermal experience
6	How do you feel in terms of thermal perception? Very hot/hot/slightly warm/neutral/slightly cool/cold/very cold (7-point scale from $+3$ to -3 in accordance with the ASHRAE thermal sensation scale) • To understand the subject's thermal sensation
7	Is the subject's head/body exposed to direct sunlight? (observation by interviewer) Yes/No • To understand if the subject is exposed to direct sun at the time of the survey
8	Overall, what would you say about this place? Reason to understand the subject's perception of the overall comfort Very comfortable/comfortable/uncomfortable/very uncomfortable (4-point scale from $+2$ to -2) • To understand the subject's perception of the overall comfort

(Fig. 8.5). Clothing insulation, expressed in the unit clo, is defined as the resistance to sensible heat transfer provided by a clothing ensemble, as the sum of individual garment clothing value.

8.4 Outdoor Thermal Comfort Standard in Hong Kong

The purpose of the thermal comfort survey was to determine the requirements for air ventilation for pedestrians in Hong Kong. As PET is defined by a number of microclimatic variables, the data obtained from the thermal comfort survey were analysed to attain the understandings based on the microclimatic conditions when

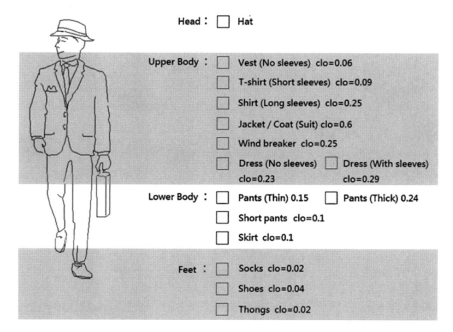

Fig. 8.5 Garment checklist was extracted from ASHRAE Standard 55-2010 and ISO Standard 7730

the survey was conducted. As suggested in Ng and Cheng (2012), the neutral PET for typical summer conditions in Hong Kong is 28.1 °C for outdoor temperature air temperature of 27.9 °C (mean summer temperature of Hong Kong, Fig. 8.6). Using the PET model, a set of possible microclimatic conditions were developed for this neutral temperature (Table 8.2), assuming that the outdoor air temperature of 27.9 °C and relative humidity of 80% (typical summer conditions in Hong Kong). Results show that higher wind speed is required to compensate for the radiant heat load in order to maintain the neutral PET in the outdoor environment. It indicates that for a person standing or walking under shade (mean radiant temperature of 32–34 °C) under typical summer conditions (air temperature of 27.9 °C), a light breeze of 0.9–1.3 m/s is required in order to achieve the neutral thermal sensation.

Findings of the survey also show that the chance of obtaining TS = 0 would be greater when lower air temperature and higher wind speed are combined. It shows that the percentage of the respondents getting TS = 0 increases gradually with an increasing wind speed (Fig. 8.7). As such, there are a number of possible strategies of improving the wind environment and thermal comfort in outdoor environments in the summer:

- To provide a conducive wind environment with wind speeds of 0.53–1.30 m/s in the city through better spatial planning and optimised potential urban development, through building coverage, layout, and disposition;

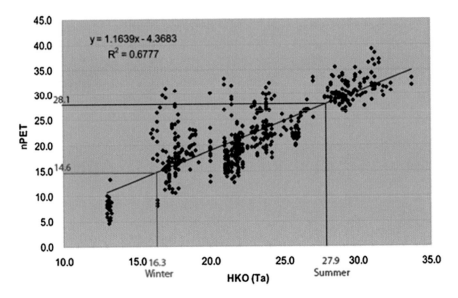

Fig. 8.6 Scatter plot of neutral PET and air temperature recorded at Hong Kong Observatory Headquarter station (Ng and Cheng 2012)

Table 8.2 Microclimatic conditions for neutral PET of 28.1 °C	Air temperature (°C)	Mean radiant temperature (°C)	Wind speed (m/s)
	27.9	28	0.20
		30	0.53
		32	0.87
		34	1.30
		36	1.76
		38	2.26
		40	2.83
		42	3.51
		44	4.08

Physiological equivalent temperature = 28.1 °C, Relative humidity = 80%, Clothing = 0.3

- To reduce radiative gains by pedestrians in streets or open spaces by providing sufficient shading opportunities, such as canopies covering building recesses and walkways, as well as colonnades. T_{mrt} under shades is slightly higher than air temperature. For example, a summer temperature of 27.9 °C corresponds to T_{mrt} in the shade of approximately 30–34 °C, whereas T_{mrt} can reach up to 50–60 °C under direct sun exposure; and
- To reduce the localised thermal load through the provision of greening, including trees and their canopies, shrubs, flower beds, and grass areas. The evapotranspiration of plants reduces sensible heat flux and hence air temperature. Urban green

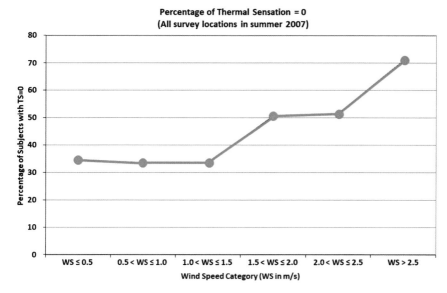

Fig. 8.7 Scatter plot of neutral PET and air temperature recorded at Hong Kong Observatory Headquarter station (Ng and Cheng 2012)

space with a size of approximately 100×100 m results in lower temperatures of 2–3 °C than in surrounding streets. Such urban oases enhance the thermal conditions of pedestrian environment for urban dwellers in Hong Kong.

8.5 Collating the Urban Climatic Map

Previous studies suggested that a 1 °C rise in T_a corresponds to an increase in PET by 1 °C. However, PET is also found to be inversely proportional to wind speed so that an increase of wind speed from 0.5 to 1.5 ms^{-1} decreases PET by at most 2 °C. According to Ng and Cheng (2012), the neutral PET value (i.e. human body feels neither cool nor warm) for summer conditions in Hong Kong is 28.1 °C. It provides the basis for the effect of the six parameters (layers) on outdoor thermal comfort and subsequently defines the classification values in the UC-AnMap. Increases or decreases in PET due to the magnitude of an urban morphological parameter allow for a balanced and synergetic consideration formulating the UC-AnMap when all the parameters are collated. It is assumed that a 1-class increase in the parameter would result in an increase in PET by at most 1 °C. Taking building volume as an example, the variation in thermal load due to different volumetric heat capacity of building can result in an increase of PET by up to 4 °C. Therefore, this parameter would have four positive categories from 1 to 4. The two important aspects, namely thermal load and dynamic potential, are collectively considered, by adopting PET as the synergistic element to collate the six layers of the UC-AnMap (Fig. 8.8).

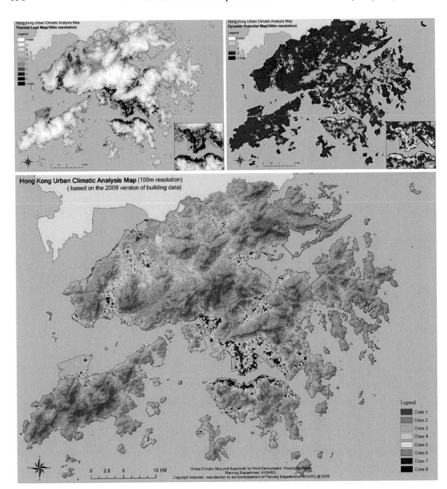

Fig. 8.8 Layers of thermal load (top left) and dynamic potential (top right) constituting the urban climatic analysis map (bottom)

The six basic layers represent three aspects which influence urban climate, including building, land use, and topography. Land use and cover information is generally used as the basic input of UCMap under low-density urban scenario (Mayer 1988; Ren et al. 2011). The mixed land use in high-density cities like Hong Kong requires additional information in order to more accurately define the climatopes (Ng 2009). In the UC-AnMap of Hong Kong, building volume and ground coverage are calculated from urban morphological such as building towers, podiums, street patterns, and open space in order to quantify urban density which affects surface roughness and heat capacity of urban areas (Ng et al. 2008). The fine resolution (100 m grid) can also capture the spatial heterogeneities of urban climatic conditions within a district so that problematic areas can be identified for urban planners and designers to take actions.

Two major components of the UC-AnMap, namely thermal load and dynamic potential, were evaluated for their corresponding implications on the urban climatic conditions of Hong Kong. The urban climatic classes were defined according to the effect of these two components (i.e. six basic layers) on outdoor thermal comfort under summer daytime conditions since it is the most critical situation in terms of thermal comfort and heat stress in Hong Kong (Ng and Cheng 2012). The wind information layer does not only provide information about the prevailing wind conditions but also identifies localised air movements such as land and sea breezes, katabatic wind, and channelling effects. Such information allows urban designers to make efficient use of localised air movements to mitigate the high thermal load in particular areas of the city.

8.6 Conclusions

The UC-AnMap of Hong Kong was developed based on six basic layers with an additional wind information layer. It describes the spatial variation of urban climatic conditions of the city in the context of outdoor thermal comfort under summer daytime conditions. It also summarises the scientific understandings of the effects of various urban morphological parameters on the urban climate which allows urban planners and designers as well as policymakers to better understand the urban climatic environment of the city. Therefore, mitigation measures and action plans can be developed in order to reduce anthropogenic heat release, improve air ventilation, reduce thermal load by shading and greenery, and creating or preserving air paths.

References

Alcoforado, M.J., H. Andrade, A. Lopes, and J. Vasconcelos. 2009. Application of climatic guidelines to urban planning: The example of Lisbon (Portugal). *Landscape and Urban Planning* 90: 56–65.

Bitan, A. 1988. The methodology of applied climatology in planning and building. *Energy and Buildings* 11: 1–10.

de Schiller, S., and J.M. Evans. 1990–1991. Bridging the gap between climate and design at the urban and building scale: Research and application. *Energy and Buildings* 15–16: 51–55.

Eliasson, I. 2000. The use of climate knowledge in urban planning. *Landscape and Urban Planning* 48: 31–44.

Gagge, A.P., and R.R. Gonzalez. 1974. Physiological and physical factors associated with warm discomfort in sedentary man. *Environmental Research* 7: 230–242.

Humphreys, M.A. 1977. The optimum diameter for a globe thermometer for use indoors. *Building Research Establishment Current Paper* 78 (9): 1–5.

ISO Standard 7730. 1994. *Moderate Thermal Environments—Determination of the PMV and PPD Indices and Specification of the Conditions for Thermal Comfort*. International Organization for Standardization.

Lindberg, F. 2004. Microclimate and behaviour studies in an urban space. In *Proceedings of Public Space Conference*, Lund, Sweden.

Matzarakis, A. 2005. Country report: Urban climate research in Germany. *IAUC Newsletter* 11: 4–6.

Mayer, H. 1988. Results from the research program "STADTKLIMA BAYERN" for urban planning. *Energy and Buildings* 11 (1–3): 115–121.

Mills, G. 1997. An urban canopy-layer climate model. *Theoretical and Applied Climatology* 57: 229–244.

Mills, G. 2006. Progress toward sustainable settlements: A role for urban climatology. *Theoretical and Applied Climatology* 84: 69–76.

Mills, G., H. Cleugh, R. Emmanuel, W. Endlicher, E. Erell, G. McGranahan, E. Ng, A. Nickson, J. Rosenthal, and K. Steemer. 2010. Climate information for improved planning and management of mega cities (needs perspective). *Procedia Environmental Sciences* 1: 228–246.

Nagara, K. 1996. Evaluation of the thermal environment in an outdoor pedestrian space. *Atmospheric Environment* 30: 497–505.

Ng, E., and V. Cheng. 2012. Urban human thermal comfort in hot and humid Hong Kong. *Energy and Buildings* 55: 51–65.

Ng, E., L. Katzschner, Y. Wang, C. Ren, and L. Chen. 2008. *Working Paper No. 1A: Draft Urban Climatic Analysis Map. Urban Climatic Map and Standards for Wind Environment—Feasibility Study*. Technical Report for HKSAR Planning Department. Hong Kong; HKSAR Planning Department.

Ng, E., 2009. *Designing High-Density Cities for Social and Environmental Sustainability*. London, Sterling, VA: Earthscan.

Nichol, J., and C.M. Lee. 2005. Urban vegetation monitoring in Hong Kong using high resolution multispectral images. *International Journal of Remote Sensing* 26 (5): 903–918.

Nikolopoulou, M., N. Baker, and K. Steemers. 1999. Improvements to the globe thermometer for outdoor use. *Architectural Science Review* 42: 27–34.

Oke, T.R. 1987. *Boundary Layer Climates*. London: Routledge.

Penwarden, A.D. 1973. Acceptable wind speeds in towns. *Building Science* 8: 259–267.

Ren, C., E.Y.Y. Ng, and L. Katzschner. 2011. Urban climatic map studies: A review. *International Journal of Climatology* 31 (15): 2213–2233.

Ren, C., K.L. Lau, K.P. Yiu, and E. Ng. 2013. The application of urban climatic mapping to the urban planning of high-density cities: The case of Kaohsiung, Taiwan. *Cities* 31: 1–16.

Sasaki, R., M. Yamada, Y. Uematsu, and H. Saeki. 2000. Comfort environment assessment based on bodily sensation in open air: Relationship between comfort sensation and meteorological factors. *Journal of Wind Engineering & Industrial Aerodynamics* 87: 93–110.

Scherer, D., U. Fehrenbach, H.-D. Beha, and E. Parlow. 1999. Improved concepts and methods in analysis and evaluation of the urban climate for optimizing urban planning process. *Atmospheric Environment* 33: 4185–4193.

Tacken, M. 1989. A comfortable wind climate for outdoor relaxation in urban areas. *Building and Environment* 24: 321–324.

Tokyo Metropolitan Government. 2005. *Thermal Environment Map for Tokyo*.

Wu, M.C., Y.K. Leung, W.M. Lui, and T.C. Lee, 2008. *A Study on the Difference Between Urban and Rural Climate in Hong Kong*. Hong Kong Observatory Reprint No. 745. Hong Kong: Hong Kong Observatory.

Yim, S.H.L., J.C.H. Fung, A.K.H. Lau, and S.C. Kot. 2007. Developing a high-resolution wind map for a complex terrain with a coupled MM5/CALMET system. *Journal of Geophysical Research* 112: D05106.

Epilogue

Throughout my research career, I learned that outdoor thermal comfort research predominantly focuses on the relationship between human subjective perception and the objective surrounding environment that city dwellers are exposed. However, this is not a constant relationship as we see from the various and sometimes contradictory results in our studies. This raises a fundamental question of whether these are due to peoples' misconceptions or the variability in peoples' perceptions. As the outdoor environment, especially in high-density cities, is constantly changing, there are still a lot of uncertainties in this field.

City dwellers are often treated as passive recipients of the environment. However, we often forget that human beings have survived on our planet Earth by successfully adapting to changing environments, at least so far. Urban planners and designers need to rethink how we can improve our living environment using more adaptive approaches. We should not underestimate humans' potential in the pursuit of comfort, health and well-being. In this sense, the focus should not be on creating the best environment, but on providing the most opportunities for city dwellers to adapt instead.

In her book, *The Death and Life of Great American Cities,* Jane Jacob advocated the need for diversity and against the overly dominant mode of living. This applies to the environment that we are exposed to in our daily life—streets, squares, parks, shopping malls, resting places, all you can name. Vibrant cities encompass a plurality of uses with constantly changing activities throughout different times. Urban inhabitants can therefore find their own ways of living and more importantly, enjoy their living.

Lisa Heschong, in her book *Thermal Delights in Architecture,* speculated that humans' environmental receptors potentially play a crucial role in thermal perception. This extends the understanding of thermal comfort to another dimension, yet it is in line with Jane Jacob's argument about the diversity of cities. Interactions between different environmental stimuli imply the need for more comprehensive approaches

K. K.-L. Lau et al., *Outdoor Thermal Comfort in Urban Environment,*
SpringerBriefs in Architectural Design and Technology,
https://doi.org/10.1007/978-981-16-5245-5

to urban planning and design, which require interdisciplinary efforts. Social scientists and psychologists offer insights into people's behaviour and how their thermal perception is influenced by environmental stimuli. As such, the design of outdoor spaces should not just be a "design" by urban designers, but also a "masterpiece" by the community.

I envisage that the future of cities lies in peoples' own hands—they can create cities in a way that they like them to be, instead of being given the cities planned and designed by someone else. Quoting my teacher, "Our future generations shall not spend their lives correcting the mistakes that we made in the past". I strongly believe that the future of cities is in their own hands.

Kevin Lau, February 2021.

Printed in the United States
by Baker & Taylor Publisher Services